NUREG/CP-0193

Proceedings of the International Workshop on Conceptual Model Development for Subsurface Reactive Transport Modeling of Inorganic Contaminants, Radionuclides, and Nutrients

Manuscript Completed: November 2005
Date Published: March 2006

Prepared by:
R. T. Cygan, M. D. Siegel, and L. J. Criscenti

Sandia National Laboratories
Albuquerque, NM 87185-0754

Edward O'Donnell, NRC Project Manager

Prepared for:
Division of Systems Analysis and
Regulatory Effectiveness
Office of Nuclear Regulatory Research
U.S. Nuclear Regulatory Commission
Washington, DC 20555-0001

ABSTRACT

An international workshop on *Conceptual Model Development for Subsurface Reactive Transport Modeling of Inorganic Contaminants, Radionuclides, and Nutrients* was held April 20-24, 2004 at La Posada de Albuquerque in Albuquerque, New Mexico. The workshop was organized and convened by the Federal Working Group on Subsurface Reactive Solute Transport Modeling and sponsored by the Federal Interagency Steering Committee on Multimedia Environmental Models (ISCMEM). The workshop themes were organized by processes: physical properties/coupling, sorption, precipitation/dissolution, and redox/microbiology. The workshop objectives were to 1) evaluate the state-of-the-art in conceptual model development and parameterization for reactive transport modeling as applied to soil, vadose zone, and groundwater systems and 2) prioritize research directions that would enhance the practical utility of reactive transport modeling. Plenary presentations and discussions provided an overview of critical modeling issues and needs, various agency applications, specifics of the four primary themes, and examples of complex applications of reactive transport models. Breakout discussion groups generated reports summarizing the motivation, current state-of-the-science, expectations, and recommendations for research in each of the four theme topics.

FOREWORD

Sandia National Laboratories prepared these conference proceedings under an Interagency Work Order with the U.S. Nuclear Regulatory Commission (NRC), and in collaboration with the Federal Working Group on Subsurface Reactive Solute Transport Modeling sponsored by the Federal Interagency Steering Committee on Multimedia Environmental Models (ISCMEM)[1]. ISCMEM consists of nine U.S. Federal agencies working cooperatively under a formal memorandum of understanding, with the objective of facilitating collaboration among participating Federal agencies in all aspects of research and development regarding multimedia environmental models.

These proceedings summarize the presentations and discussions at an international workshop on reactive transport modeling held in Albuquerque, New Mexico, in April 2004. Specifically, the presentations and discussions focused on various approaches for evaluating the fate of inorganic contaminants including radionuclides, which are of particular interest to the NRC. Abstracts and papers associated with the presentations are included, along with summary reports for the four breakout sessions of the workshop, which were organized to assess: (1) coupling of physical and reactive processes, (2) sorption processes, (3) nucleation, precipitation and dissolution processes, and (4) redox (reduction-oxidation) and biologically mediated processes. These conference proceedings are significant because they address issues that are of concern wherever a Federal agency evaluates the potential effects of environmental contamination attributable to its actions. The models employed are complex and evolving, and it is important for Federal agencies to share their information and analytical techniques to promote consistent results in the assessment of the related systems.

The views and opinions presented in this report and, in particular, the recommendations put forth in the breakout summaries, are those of the individual participants, and publication of this report does not necessarily constitute NRC approval or agreement with the information contained herein. As such, these proceedings are not a substitute for NRC regulations. Rather, the approaches and methods described in these proceedings and the recommendations from the breakout sessions are provided for information only, and compliance is not required. Moreover, use of product or trade names herein is for identification purposes only and does not constitute endorsement by the NRC or Sandia National Laboratories.

Carl J. Paperiello, Director
Office of Nuclear Regulatory Research
U.S. Nuclear Regulatory Commission

---

[1]A modified version of these proceedings has been published on the ISCMEM web site at www.iscmem.org.

CONTENTS

ACKNOWLEDGEMENTS

The authors are grateful for the support of the United States Nuclear Regulatory Commission and the valuable assistance provided by the project manager Edward O'Donnell. Sandia was tasked by the NRC with organizing and hosting the workshop as part its participation in the Federal Working Group on Subsurface Reactive Solute Transport Modeling sponsored by the Federal Interagency Steering Committee on Multimedia Environmental Models. We are grateful for the support provided by the scientists and engineers of this working group. In particular, we appreciate the leadership and guidance provided by James Davis of the U.S. Geological Survey who chaired and coordinated the efforts of this large group of expert researchers. We would also like to extend our gratitude to the local coordinating committee and the excellent support provided by Goldie Piatt, Mona Aragon, and Barbara Martens. We also appreciate the efforts of the organizational committee and breakout session chairs in bringing together an excellent group of leading scientists and engineers, and in designing an outstanding technical program. We are indebted to all of the participants and speakers of the workshop who contributed to the lively discussions, summary reports for the breakout sessions, and workshop proceedings. Lastly, we acknowledge the support provided by the sponsoring government agencies— Agricultural Research Services (USDA), National Oceanic and Atmospheric Administration, Natural Resources Conservation Service (USDA), U.S. Army Corps of Engineers, U.S. Department of Energy, U.S. Environmental Protection Agency, U.S. Geological Survey, and, of course, the U.S. Nuclear Regulatory Commission.

Sandia is a multi program laboratory operated by Sandia Corporation, a Lockheed Martin Company, for the United States Department of Energy's National Nuclear Security Administration under Contract DE-AC04-94AL85000.

1. INTRODUCTION

An international workshop on Conceptual Model Development for Subsurface Reactive Transport Modeling of Inorganic Contaminants, Radionuclides, and Nutrients was held April 20-24, 2004 at La Posada de Albuquerque in Albuquerque, New Mexico. The workshop was organized and convened by the Federal Working Group on Subsurface Reactive Solute Transport Modeling and sponsored by the Federal Interagency Steering Committee on Multimedia Environmental Models (ISCMEM). The primary intent of the workshop was to draw on the expertise of a variety of scientists from government agencies and leading universities to better assess the state of the art in reactive transport modeling, and to identify critical areas for conceptual model development and advances in scientific understanding. Emphasis was placed on the modeling of inorganic contaminants, including radionuclides, and in examining the influence of nutrients on reactive transport. Seventy-four experts representing government laboratories, research universities, and private industries attended the workshop (see Section 5.).

The workshop was organized along two dimensions: applications and processes. The applications were introduced during the first day and were followed throughout the workshop (see Section 2). The technical review sessions were organized by processes: Physical/Coupling; Sorption; Redox/Microbiology; Precipitation/Dissolution. Breakout sessions were organized by the same topics to allow more detailed discussions and identification of critical scientific needs. During the breakout sessions, we also identified research projects on the corresponding process that are directly relevant to agency applications. The final plenary session allowed for all these issues to be pulled together—by listing the research topics that will provide the greatest benefit to the most applications. Findings from each of the breakout discussion groups are presented as summary reports (see Section 3). Proceedings reports and abstracts for the oral presentations are provided in Section 4.

A report summarizing the general findings of the workshop was published in a news publication of the American Geophysical Union (Davis et al., 2004, *Eos, Transactions of the American Geophysical Union*, 85(44), 449-455). Additionally, electronic web-based proceedings are published on the public web pages of the Interagency Steering Committee on Multimedia Environmental Models (www.iscmem.org). Electronic versions of the workshop presentations are also available through the ISCMEM web site.

The following individuals helped with the planning and organization of the workshop:

Local coordination committee

Malcolm D. Siegel	Sandia National Laboratories
James A. Davis	U.S. Geological Survey
Louise J. Criscenti	Sandia National Laboratories
Goldie Piatt	Sandia National Laboratories
Mona Aragon	Sandia National Laboratories
Randall T. Cygan	Sandia National Laboratories

Interagency organizational committee

John A. Apps	Lawrence Berkeley National Laboratory
Alex E. Blum	U.S. Geological Survey
Louise J. Criscenti	Sandia National Laboratories
James A. Davis	U.S. Geological Survey
Sabine R. Goldberg	George E. Brown Jr. Salinity Laboratory
Kenneth M. Krupka	Pacific Northwest National Laboratory
Robert W. Puls	National Risk Management Research Laboratory, USEPA
Malcolm D. Siegel	Sandia National Laboratories
Carl I. Steefel	Lawrence Berkeley National Laboratory
David R. Turner	Southwest Research Institute
Steve Yabusaki	Pacific Northwest National Laboratory
John M. Zachara	Pacific Northwest National Laboratory

2. WORKSHOP AGENDA

Conceptual Model Development for Subsurface Reactive Transport Modeling of Inorganic Contaminants, Radionuclides, and Nutrients

La Posada de Albuquerque
Albuquerque, New Mexico
April 19-22, 2004

Monday, April 19, 2004	
6:00-7:00 pm	Registration (Mezzanine) Icebreaker (Lobby) (cash bar and hors d'oeuvres)

Tuesday, April 20, 2004	
7:00 am	Registration (Mezzanine) Continental Breakfast (Eulalia's Restaurant)
8:00 am	Opening Session (Ballroom)
8:00 am	Workshop Logistics (Malcolm Siegel - Sandia National Laboratories)
8:10 am	Welcoming Remarks (Jim Davis - U.S. Geological Survey)
8:20 am	**Plenary Session 1–Overview of Modeling Issues** (Ballroom) Session Leaders: George Yeh (UCF) and Carla Koretsky (WMU)
8:20 am	Geochemistry, Groundwater and Pollution: Learning by Modeling (Tony Appelo - Hydrochemical Consultant)
8:50 am	Reactive-Transport Prediction Uncertainty and Simulation Accuracy, Observation Errors, and Sensitivity Analysis (Mary Hill - U.S. Geological Survey)
9:20 am	Use of Reactive-Transport Models in Field Studies: Experience with the PHAST Simulator (David Parkhurst - U.S. Geological Survey)
9:50 am	Coffee Break (Mezzanine)
10:05 am	**Plenary Session 2–Overview of Agency Applications and Modeling Needs** (Ballroom) Session Leaders: Malcolm Siegel (SNL) and Ken Krupka (PNNL)
10:05 am	EPA Modeling Applications and Needs (Robert Puls – U.S. Environmental Protection Agency)
10:35 am	Reactive Transport Modeling of the Yucca mountain Site, Nevada (Bo Bovardsson - Department of Energy)
10:55 am	Conceptual Model Development, Remediation, and Reactive Transport Modeling at Hanford: The U.S. DOE's Largest Legacy Waste Site (John Zachara - Pacific Northwest National Laboratory)
11:15 am	NRC Modeling Applications and Needs (William Ott – Nuclear Regulatory Commission)
11:30 pm	Predicting Soil and Water Chemistry in and Below the Root Zone of Agricultural Lands: Major Ions, Nutrients and Toxic Ions (Don Suarez - Agricultural Research Service)
11:45 pm	ERDC/USACE Modeling Applications and Needs (Chris McGrath - Engineer Research and Development Center)
NOON	Lunch (Eulalia's Restaurant)
1:00 pm	**General Discussion** (Ballroom) Discussion leader: Louise Criscenti (SNL)

1:20 pm	**Plenary Session 3–Physical Properties and Coupling with Reactive Processes** (Ballroom) Session Leaders: Caroline Stevens (US EPA) and Andy Tompson (LLNL)
1:20 pm	Direct Visualization of Reactive Transport: Mixing at the Pore-Scale and at the Darcy-Scale (Charles Harvey - Massachusetts Institute of Technology)
1:50 pm	Modeling Fate and Transport of Inorganic-Organic Co-contaminants at a DOE Fractured Waste Site (Jack Gwo - University Maryland Baltimore County)
2:20 pm	Discussion of Plenary Session 3; discussion of scaling issues - How are the physical properties related to the conceptual model "gap" for reactive processes?
2:40 pm	**Plenary Session 4–Sorption Processes** (Ballroom) Session Leaders: Bruce Honeyman (CSM) and Janet Hering (CalTech)
2:40 pm	Radionuclide Sorption in High-Level Waste Performance Assessment: Abstraction of Results from Experiments and Surface-Complexation Models (Roberto Pabalan – Southwest Research Institute)
3:10 pm	Coffee Break (Mezzanine)
3:25 pm	**Plenary Session 4–Sorption Processes continued** (Ballroom)
3:25 pm	Application of Surface Complexation Modeling to Describe Uranium(VI) Adsorption and Retardation at the Field Scale (Gary Curtis - U.S. Geological Survey)
3:55 pm	Reactive Transport Modeling of Multicomponent Ion Exchange at the Laboratory and Field Scale (Carl Steefel - Lawrence Livermore National Laboratory)
4:25 pm	Discussion of Plenary Session 4 - Gap analysis
4:50 pm	**Plenary Session 5–Discussion of Breakout Session Objectives** (Ballroom) Session Leaders: Jim Davis (USGS) and John Zachara (PNNL)
5:05 pm	Begin Breakout Sessions
	Working Group #1–Physical Properties and Coupling with Reactive Processes Discussion Leaders: Carl Steefel (LLNL) and Steve Yabusaki (PNNL) Plaza Suite **Working Group #2–Sorption Processes** Discussion Leaders: Louise Criscenti (SNL) and Jim Davis (USGS) Boardroom **Working Group #3–Nucleation, Precipitation, and Dissolution Processes** Discussion Leaders: John Apps (LBNL), Alex Blum (USGS), and Ken Krupka (PNNL) Continental Room **Working Group #4–Redox Processes and Biologically Mediated Processes** Discussion Leaders: Gary Curtis (USGS) and John Zachara (PNNL) Ballroom
6:00 pm	Adjourn

Wednesday, April 21, 2004	
7:00 am	Hot Breakfast Buffet (Eulalia's Restaurant)
8:00 am	Opening Remarks/Logistics (Ballroom)
8:10 am	**Plenary Session 6–Nucleation, Precipitation, and Dissolution Processes** (Ballroom) Session Leaders: Sue Brantley (PSU) and Janet Herman (UVA)
8:10 am	Conceptual Approaches for Scaling from Molecular to Macroscopic Levels of Nucleation and Precipitation (Nita Sahai - University of Wisconsin - Madison)
8:35 am	Determining Dissolution, Precipitation and Nucleation Rate Laws in Natural Systems (Alex Blum - U.S. Geological Survey)
9:00 am	Precipitation-Front Modeling: Issues Relating to Nucleation and Metastable Precipitation in the Planned Nuclear Waste Repository at Yucca Mountain, Nevada (Eric Sonnenthal - Lawrence Berkeley National Laboratory, John Apps - Lawrence Berkeley National Laboratory)
9:30 am	Discussion of Plenary Session 6 - Gap analysis
9:55 am	Coffee Break (Mezzanine)
10:10 am	**Plenary Session 7–Redox Processes and Biologically-Mediated Processes** (Ballroom) Session Leaders: Dave Blowes (UW) and Tim Ginn (UC-Davis)
10:10 am	Reactive Transport Modeling of Biologic Iron(III) Reduction Under Varied Hydrologic Conditions (Bill Burgos - Penn State University)
10:40 am	Surface Chemical and Thermodynamic Controls on Bacterial Metal Reduction in Subsurface Environments (Eric Roden - University of Alabama)
11:10 am	In-situ Biogeochemical Rate Laws in Sediments and Aquifers (Phillippe Van Capellen - Utrecht University)
11:40 am	Discussion of Plenary Session 7 - Gap analysis
12:05 pm	Lunch (Eulalia's Restaurant)
1:00 pm	**Plenary Session 8–Complex Applications** (Ballroom) Session Leaders: John Westall (OSU) and Al Valocchi (UIUC)
1:00 pm	Testing Coupled Microbial and Reactive Solute Transport Models with In-situ Experiments: REX and Redox Zone Experiments at Aspo (Sweden) (Javier Samper - University of La Coruna, Spain)
1:30 pm	Modeling Thermal-Hydrologic-Chemical (THC) Coupled Processes with Application to Underground Nuclear Tests at the Nevada Test Site: A "Grand Challenge" Supercomputing Problem (Peter Lichtner - Los Alamos National Laboratory)
2:00 pm	Challenges in Modeling Evolving Reactivity – Examples from Groundwater Contamination and Remediation Studies (Uli Mayer - University of British Columbia)
2:30 pm	Discussion of Plenary Session 8 - Gap analysis
2:50 pm	**General Discussion** Reminder of workshop and breakout session objectives; discussion of breakout session progress from yesterday Discussion Leaders: Carl Steefel (LBNL) and John Apps (LBNL)
3:15 pm	Coffee Break (Mezzanine)
3:30 pm	Continue Breakout Sessions
	Working Group #1 - Plaza Suite Working Group #2 - Boardroom Working Group #3 - Continental Room Working Group #4 - Ballroom
6:00 pm	Adjourn
7:00 pm	Shuttle to Bien Shur Restaurant leaves from hotel loading zone on Copper Street

Thursday, April 22, 2004	
7:00 am	Health Nut Continental Breakfast (Mezzanine)
8:00 am	Opening remarks/logistics/discussion of the breakout session progress and the products needed from the breakout sessions (Ballroom) Discussion Leaders: Jim Davis (USGS) and Louise Criscenti (SNL)
8:20 am	Continue Breakout Sessions
	Working Group #1 - Plaza Suite Working Group #2 - Boardroom Working Group #3 - Continental Room Working Group #4 - Ballroom
10:00 am	Coffee Break (Mezzanine)
10:15 am	**Plenary Session 9–Spanning the Gap: Integration of New Research and Application Ideas for Reactive Transport Modeling at the Field Scale** (Ballroom) Session Leaders: Jim Davis (USGS), John Zachara (PNNL), Carl Steefel (LBNL), and John Apps (LBNL)
12:30 pm	Closing Lunch (Eulalia's Restaurant)
1:30 pm	Meetings of breakout session chairs and session scribes (Ballroom)
4:00 pm	Adjournment

3. BREAKOUT SESSION SUMMARIES

3.1 Coupling of Physical and Reactive Processes

Steve Yabusaki and Carl I. Steefel

3.1.1 Motivation
In many subsurface situations of consequence (e.g., contaminant hydrogeology, petroleum extraction, carbon sequestration, etc.), scientists and engineers are being asked by decision-makers to predict the fate of components of concern at the field scale under conditions where both reactions **and** transport are processes of first-order importance. These situations and the modeling tools and approaches that are used to analyze them were the subject of a workshop, "Conceptual Model Development for Subsurface Reactive Transport Modeling of Inorganic Contaminants, Radionuclides, and Nutrients," held at the La Posada in Albuquerque, New Mexico on April 20-22, 2004.

Discussions from the breakout sessions on "Coupling of Physical and Reactive Processes" are summarized here. The participants in this group addressed issues related to the fate of reactive solutes in complex field systems, where spatially variable material properties directly influence not only the physical processes of flow and transport, but also the biogeochemical reactions and rates. Of particular interest was the interplay between physical and reaction processes, and how this coupling could be efficiently and realistically accounted for in reactive transport models.

3.1.2 Background
Historically, the modeling of reactions and transport processes at the field scale were seldom treated with equal importance. This was a direct legacy of the background of the modelers, who were typically trained in hydrology rather than biogeochemistry. Hydrologically-oriented, field-scale flow and transport models addressed complexity primarily through the specification of heterogeneous, spatially-distributed hydrologic properties that reproduced the spatial and temporal response of the aquifer potential surface to transient stresses. Chemical reactions were included only to the degree that they could be easily incorporated into the existing transport formulations: a single component with typically linear decay and/or infinite capacitance linear isotherms. The simplified chemical treatment expressly omitted reactions between chemical components, i.e., multicomponent effects.

On the other hand, the modeling of multicomponent geochemical reactions based on standardized thermodynamic databases historically focused on complexity in the chemical reactions and rate laws at the laboratory scale rather than complex field-scale transport situations. Materials were typically homogenized and studied in batch settings or, more rarely, in columns where constant and uniform flow conditions were imposed.

As scientists and engineers respond to the current demand for predictive understanding of field-scale reactive transport, it is clear that heterogeneity and scale issues are a formidable scientific challenge. In this respect, the field-scale representation of reactivity is in its infancy, especially where flow and transport processes are limiting the 1) mixing of reactive components and 2) access to reactive surfaces. For example, preferential flow paths and exclusion zones resulting from multiscale physical and chemical heterogeneities can create opportunities for disequilibrium between adjacent regions. The situation can be further complicated by reactions that alter material properties controlling flow and transport, and/or the reactivity of mineral surfaces.

3.1.3 Contemporary Issues

A principal difficulty for developing defensible predictive reactive transport simulators is the identification of appropriate field-scale process models in the presence of multi-scale heterogeneities. In terms of physical properties, spatially-variable permeability and porosity can result in preferential flow paths that dictate supply rates and residence times of transported components. Non-uniform distributions of reactive components arising from these conditions are subject to diffusive mixing enhanced by dispersion processes that are linked directly to the heterogeneous character of the pore network. From the reactive transport perspective, these mixing processes provide the driving force for intra-aqueous reactions between solutions that are not in chemical equilibrium. While these reactions may be fast, the rate-limiting step of the reaction may be determined by kinetically-controlled mass transfer between adjacent zones with differing chemistries. Examples of this kind of behavior can be found at all length scales, including interactions between (1) fracture and matrix zones (dual porosity), (2) macropores and structured soils; (3) mobile and immobile water (multi-region), and (3) primary and secondary porosity.

Multi-scale heterogeneities in chemical properties are typically associated with spatially variable mineral assemblages. In this case, mineral reactions can create non-uniform component distributions even in the absence of physical heterogeneities. For reactive transport models that include heterogeneous (aqueous-solid) phase reactions, reactive surface area is a critical parameter controlling the sites and rates of reaction. However, the reactive surface area of individual minerals/phases that are required by mechanistic mineral reactions and rate laws is very difficult to measure in natural systems. Standard gas-adsorption BET surface area measurements are less than adequate because the total surface area measured is not apportioned to individual minerals and does not distinguish the fraction that is hydrologically accessible and available for reaction. Spatial variability and uncertainty in the characterization of these specific attributes of the surface area further compounds the complexity originating in the mineral assemblage itself.

In light of these issues, the challenge is to develop field-scale parameterizations that appropriately address the combined impact of physical and chemical heterogeneities on flow, transport, and reaction processes. Sufficiently detailed resolution of flow and transport is needed to capture the range of transport-limited to reaction rate-limited behaviors exhibited by spatially distributed components interacting with local distributions of reactive surfaces. Furthermore, small-scale processes and properties below the resolution of computational grids must be addressed if they are important to the behaviors at the modeling scale. Examples of upscaled (effective) parameterizations for sub-grid variability in hydrologic and biogeochemical processes include dispersivity, multi-region mass transfer, and multiple sorption site types. One reason to consider this level of modeling detail is to adequately represent the fingering and homogenizing behavior that results from commonly observed spatial correlations between physical and chemical heterogeneities.

Conversely, there are important field-scale processes (e.g., buoyancy driven flows of liquids and gases) that are not easily observed under typical laboratory experimental conditions. One situation that significantly complicates the interpretation of field-scale behavior (and thus, modeling of it) is when dynamic reactive transport processes take place in a system where the reactivity is evolving. Over time, reaction-induced changes to liquid density, liquid saturation, mineral volume, and biomass can alter velocities and transport pathways, such that reactant delivery rates and residence times are significantly affected. The Grimsel experimental site in Switzerland is a case where the injection of hyperalkaline fluids into a fracture zone caused a reduction in the overall permeability, as indicated by the higher pressure needed to maintain a constant injection rate. However, the reduction of the overall fracture porosity enhanced average linear velocities, resulting in earlier tracer arrival. The general difficulty is relating reactions and the mechanics of nucleation and/or dissolution/degradation, to material property changes that are mechanistically linked to model parameters for flow and transport.

Reactions can also directly modify solid phase reactivity. In this case, precipitation/dissolution, sorption, or biomass reactions alter the availability and/or accessibility of reactive mineral surfaces. These effects include (1) passivation of reactive surfaces, sometimes observed in permeable reactive barriers; (2) creation of reactive surfaces during the precipitation of secondary minerals or generation of biomass; and (3) dissolution of minerals that have sorbed components (e.g., release of arsenic during reductive dissolution of iron minerals).

3.1.4 Modeling Approaches
At this time, the scientific community is more interested in exploring and assessing the appropriateness of alternative conceptual process models that can be included in existing reactive transport simulators, rather than in ranking simulator codes against each other. This reflects one motivation for the workshop: recognition of the current uncertainty in the specification and selection of appropriate and scientifically-defensible conceptual process models for many field-scale situations (especially those identified above).

In the absence of mechanistic understanding at the field-scale, the consensus of the breakout session was that it was more reasonable to refine a framework of understanding (e.g., numerical representations of conceptual process models) developed from laboratory-scale experiments rather than attempting to fit global parameters to a "standard" field-scale process model without a scientific basis. Furthermore, this approach should be done with the expectation that conceptual process models and numerical parameterizations developed under idealized conditions (e.g., homogenized < 2 mm size fractions) at the laboratory scale are not likely to accurately depict the field-scale situation, which must address multiple scales of physical (e.g., hydraulic conductivity) and chemical (e.g., mineralogy) heterogeneity.

One key consideration identified by the breakout session in developing robust field-scale reactive transport simulators is the ability to test the system of coupled process models during perturbations of physical and chemical conditions that exercise as much of the relevant range of parameters as possible. In this respect, a series of history matching studies ranging from laboratory scales to the field scale would provide a stronger basis for understanding scale-up issues and the resulting field-scale predictions. The need for calibration of field-scale reaction parameters in this approach should not be underestimated. Current limitations in characterizing multiscale variability in subsurface properties and observing in situ behavior lead to knowledge gaps that may only be effectively addressed through process model calibration against observations in the field. While this may limit the robustness of model predictions, decisions based on model predictions can be justified if the conditions modeled are within the range of calibration conditions and the predicted outcomes can be bracketed to the extent that they clearly differentiate alternatives.

3.1.5 Recommendations
General areas targeted by the breakout session as being important to the advancement of field-scale reactive transport modeling included: 1) field research sites, 2) measurement technology,

and 3) computational technology. Of the three areas, considerably more energy was directed at the need for coordinated, comprehensive research dedicated to modeling field-scale behaviors at a particular site. An important requirement of this approach was the assembly of multi-disciplinary teams to adequately address the range of scales and multiple interacting processes operating at a site. This would be necessary to investigate the impact of multi-scale physical and chemical heterogeneities on the scale-up to field scale process models and parameterizations. In particular, experiments and characterization studies at complementary length scales in the field and in the laboratory were identified as critical research components. In support of these studies, geophysical imaging (< 1 mm) of density, moisture content, and chemical components was identified as an important tool in developing high-resolution data sets of in situ subsurface properties. The value of this approach lies in understanding the linkage of model processes and parameters to observable particle characteristics (e.g., pedo-transfer function approaches to estimate flow and transport parameters as well as reactive surface area).

Traditional laboratory-scale experiments and characterization studies at the batch and column scale are still fundamental components in the building of a mechanistically predictive understanding of field scale behavior. Equally important but more challenging in design and analysis are intermediate-scale experiments. These experiments performed in the field and in the laboratory provide an essential bridge from the small-scale laboratory analyses of limited dimensionality to the field-scale behaviors of interest. Both laboratory "sand box" and in situ field experiments over limited volumes are recommended. The important advantage of these experiments is the control over boundary conditions and perturbations over a limited volume of subsurface materials. Consequently, process and parameter upscaling can be studied in multi-dimensional systems with realistic degrees of freedom and variability.

With the goal of a process-oriented understanding of observed behaviors, the recommendation here is to design laboratory and in situ field experiments to explicitly include data collection for the development and testing of reactive transport models. An "early and often" approach to reactive transport modeling for predictive assessments of baseline behavior and/or engineering performance can iteratively adapt to improvements in process and property characterization while providing guidance for subsequent testing. Modeling complexity, however, must be considered in the context of overarching scientific and/or engineering objectives, and tempered by the (in)adequacy of collected data.

The investment in a comprehensive, sustained study of a field site will be considerable, requiring a thoughtful selection process. Desirable qualities in a field site identified by the breakout session attempted to strike a balance between complexity and tractability. In general, the field site should target a range of observable length scales of physical and chemical heterogeneity that allow the characterization of flow paths and initial conditions. Perturbations and propagating transients are critical to the elucidation of system response and should be observable in time and space. There should be a spectrum of measurable biogeochemical processes (e.g., mineral precipitation/dissolution, redox, aqueous and surface complexation, etc.) of interest operating in the system.

The breakout session went on to identify known situations where a desirable mix of these field site attributes was present. One example was a contaminated vadose zone and groundwater system with transient river interactions. In this case, the transients are related to both the hydrologic forcing due to changes in river stage and to the interaction between the differing chemistry of river water and groundwater. Candidate sites were described where uranium surface complexation (e.g., Hanford 300 Area, uranium mill tailings sites) and strontium ion exchange (e.g., Hanford 100-N Area) were the geochemical mechanisms of interest. Another situation of interest was drainage through mine tailings. In this case, metal contamination and low pH chemistry develops in the vadose zone. These systems include biogeochemical reactions that involve microbially-mediated oxidation of ferrous iron and sulfide minerals, and mineral weathering reactions.

While coordinated and comprehensive research programs dedicated to specific field sites are crucial to the advancement of understanding and prediction of field-scale reactive transport, it can be argued that technologies for characterization and analysis are vital to the successful achievement of that goal. Field scale reactive transport behavior is fundamentally affected by biological and geochemical heterogeneities. More work is needed to develop characterization tools for high resolution data sets that can yield relationships between material properties and biogeochemical processes and properties (e.g., identifying correlations between cation exchange capacity and geophysical measurements of grain size or porosity).

A new generation of analysis tools is necessary to process the large data sets from these high-resolution reactive transport studies. High performance computing and computational technologies should be used to full advantage by subsurface scientists and engineers in the elucidation of field-scale behaviors in the context of multi-scale physical and chemical heterogeneities. Implicit in this recommendation is the need for sustained funding to develop and maintain robust reactive transport simulators for the subsurface research and engineering community.

Acknowledgments: The authors would like to acknowledge the breakout session group that participated in the summarized discussions with special thanks to William P. Ball, Georgia Destouni, Timothy R. Ginn, James R. Hunt, K. Ulrich Mayer, and Albert J. Valocchi for their comments on this document.

3.2 Sorption Processes

James A. Davis and Louise J. Criscenti

3.2.1 Motivation
In many subsurface situations of consequence (e.g., contaminant hydrogeology, petroleum extraction, carbon sequestration, etc.), scientists and engineers are being asked by decision-makers to predict the fate of components of concern at the field scale under conditions where both reactions **and** transport are processes of first-order importance. These situations and the modeling tools and approaches that are used to analyze them were the subject of a workshop, "Conceptual Model Development for Subsurface Reactive Transport Modeling of Inorganic Contaminants, Radionuclides, and Nutrients," held at the La Posada in Albuquerque, New Mexico on April 20-22, 2004.

A major objective of the workshop was to examine and understand the "conceptual gap" between current modeling activities and conditions actually present in the field, and make recommendations as to how the gap could be closed through targeted research initiatives. In order to define research priorities for this field, the members of each breakout session were asked to evaluate the state of the art in order to define: a) advances needed in scientific understanding, b) new approaches for conceptual model development, c) improved methods for assessing the values of field-relevant reaction parameters, and d) suggestions for targeted research initiatives that would lead to new model development. Discussions from the breakout sessions on "Sorption Processes" are summarized here.

3.2.2 Technical Issues
The goal of the breakout session was to describe how a technical basis for more realistic retardation assessment can be developed and how different conceptual models for sorption processes can be evaluated and compared. A major theme of the discussion in the Adsorption-Desorption breakout session was how to move beyond the often used constant-K_d conceptual model to more realistic models that describe variable retardation by sorption processes at the field scale. The additional conceptual models considered in the discussion were surface complexation models (with conditional, field-site specific parameters), surface complexation models with unconditional constants from a standardized database, and kinetic models. The discussion involved many different issues, including:

3.2.2.1 Relationship between Site Heterogeneity and Sorption Model Uncertainty
Field sites are generally heterogeneous, but how much error is introduced by not considering the heterogeneity of the unconsolidated porous medium at the scale of hundreds of meters or tens of kilometers? Can representative field samples be collected that allow laboratory determinations of realistic K_d values or ranges of K_d values for a field site? Can criteria be developed that describe when volume averaging of porous medium properties is appropriate for a field site? Can criteria be developed that describe when multi-domain modeling is necessary (domains with different K_d values or ranges of K_d values)?

3.2.2.2 Guidelines to Assess Required Sorption Model Complexity
Recent field studies have made it clear that the transport of inorganic contaminants is significantly impacted by temporal or spatial variation of chemical conditions. When chemical conditions vary, a conceptual model for sorption that is more complex than the constant-K_d model is required for an accurate description of transport. Even when dealing with cations with simple aqueous chemistry, the constant-K_d model may not be acceptable (e.g., Lichtner et al.,

2004, *Vadose Zone Journal*, v. 3, p. 203-219). However, there are also field sites where contaminant retardation has been adequately described by the constant-K_d approach. For a specific field site, what are the criteria that indicate that either a constant-K_d conceptual model is sufficient or that a more advanced adsorption model is necessary to predict contaminant migration? What are the criteria that indicate that sorption kinetics should be considered in a predictive model for contaminant attenuation? A published document with guidelines to determine whether or not a sorption model is adequate for general conditions at a field site would be beneficial to regulators and licensees.

3.2.2.3 Selecting Appropriate Sorption Model Database

Should the conceptual model (and parameter values) for retardation at the field scale be based on laboratory data collected with soils and natural groundwater from the specific field site or on a model based on literature studies of sorption by individual mineral phases (i.e., from a compiled database)? This issue is central to the type of sorption model preferred to describe contaminant retardation in the field. The most mechanistic surface complexation models require a comprehensive database of unconditional equilibrium constants for the adsorption of all inorganic aqueous species to each soil mineral or solid. A complete database of this nature does not currently exist. This has seriously hindered the ability to determine if this type of model can be used to successfully describe contaminant adsorption under field conditions. However, surface complexation models that are parameterized based on laboratory K_d measurements with site-specific soils can describe the variability in adsorption as a function of water chemistry (e.g., pH, alkalinity) that is not described by the constant-K_d model. While such models are not mechanistic, they do allow for changes in contaminant partitioning as a function of spatially and temporally variant chemical conditions, as usually occurs in plumes.

3.2.2.4 Relationship between Data Collection and Uncertainty Reduction

All reactive transport models require field-site specific data, especially to describe groundwater or vadose zone flow. Typically, the emphasis in the collection of field data is placed on the development of a flow model for a field site. However, there is an increasing appreciation at the federal agencies that more technical justification will be needed in the future to support predictions of the outcome of Monitored Natural Attenuation (MNA). That is to say, more data collection and better predictive models of contaminant retardation will be requested by the regulators and the public. Models for MNA need to be able to account for contaminants currently in the soil and predict what will happen as chemical conditions evolve over time. With respect to parameterizing a sorption model for a field site, there is a perception that any conceptual model beyond a constant-K_d model is too costly. However, recent experience suggests that "conditional" surface complexation models capable of accounting for variable chemical conditions can be developed with additional costs that are not excessive. Field studies are needed that evaluate the tradeoff between spending more on field data collection for flow model calibration versus sorption model calibration. Such studies could determine the reduction in retardation uncertainty that results from increasing data collection for flow and sorption model parameterization.

3.2.2.5 Quantifying Model Uncertainty to Identify Knowledge Gaps

Model complexity and associated uncertainty was discussed at length. The need to have a quantitative measure of uncertainty was emphasized. The irony that using complex models that better describe the adsorption process may actually yield greater uncertainty than simpler models was addressed. The "model" uncertainty that can be quantified is clearly distinct from the "uncertainty" associated with our ignorance of adsorption mechanisms. The more we know and the more we can model, the more scenarios we can consider. While this exercise improves our understanding of possible outcomes and will result in better decision-making, it may not yield quicker resolutions regarding the future of particular field sites.

3.2.3 Panel Recommendations

3.2.3.1 Establish Field Sites for Long-Term Federal Agency Research and Model Development
The strongest recommendation is that one or more long-term field research sites should be established for conceptual model evaluation and development and comparison of reactive-transport models. These research sites should probably differ from the few that have been already studied in depth (e.g., USGS studies at Cape Cod, Massachusetts and Naturita, Colorado, see http://ma.water.usgs.gov/CapeCodToxics/default.htm and http://www.nrc.gov/reading-rm/doc-collections/nuregs/contract/cr6820/cr6820.pdf). The chosen sites should represent different archetypes in terms of soil composition, contaminant-type, etc. The sites should be instrumented for comprehensive applied research and would be chosen with the expectation that sorption was an important process controlling contaminant migration, along with other reactive processes. Long-term access to the sites would also be an important requirement.

Different sorption models should be evaluated for their ability to describe contaminant retardation at the sites, and for their ability to resolve the various technical issues described above. Detailed characterization and study of several carefully selected field sites should be used to then provide guidance on which parameters must be measured in the field and which parameters are less significant to adsorption processes or can be derived from databases. This guidance on appropriate adsorption models and data collection for a field site may vary depending on model objectives. Research at the field sites should focus on quantifying sorption occurring simultaneously with other reactive processes, evaluating the impact of substrate heterogeneity on sorption as a function of physical scale, developing better methods for measuring field-relevant reactive parameters such as desorption and isotope exchange, and providing guidance on when kinetics needs to be considered either due to aging of the contaminated field site or the relative rates of flow, adsorption and desorption. Such methods may also be useful in assessing when contaminant concentrations in groundwater are being controlled by sorption or precipitation-dissolution processes. In addition, the transferability of laboratory measured K_d values on field materials to the field system should be investigated. For added value, it is suggested that the field sites selected for the extensive research also be sites that have already been identified for MNA by a regulatory agency. Another major objective of the intensive field research is to provide a better understanding of the sensitivity of model parameters in transport simulations and how the uncertainty in model predictions may be impacted by field characterization. Along with scientific studies, an economic assessment of the cost and benefit of each laboratory or field experiment or analysis could be performed. This assessment should be incorporated into the guidelines proposed for modeling a field site.

3.2.3.2 Publication of a Guidance Document on Conceptual Models for Sorption
The second major recommendation of the panel was that a document should be published that provides a scientific framework for incorporation of sorption models into reactive-transport models. For regulatory purposes, it is important to identify field site characteristics that signal the need to use a surface complexation model rather than a constant-K_d conceptual model. It could also be demonstrated how the choice of a modeling approach is dependent upon the modeling objectives. Educational tools for the consultant community are needed to explain the benefits of using more complex sorption models, particularly if more site-specific data collection and computer resources upfront will protect the public in the long-term, hence saving their clients major expenses in the future. Examples of such documentation include reports (e.g.,

EPA Research Briefs), but some educational tools might also be distributed as one-page agency information briefs, demonstrating the difference in calculated contaminant migration using different adsorption models for an archetype field site. A decision tree about how to evaluate a site, designed for both consultants and regulators, would also benefit the community at large. Such an expert system could indicate what type of information is available to implement different adsorption models and what field site-specific information is required. Studies at the field research sites could provide data on, for example, the cost of building a field-calibrated surface complexation model and consequent reduction in uncertainty in simulated contaminant retardation.

3.2.3.3 Development of a Database for Sorption Model Parameters

There is a strong need for the improvement and standardization of a new database for sorption constants. Although considerable research has been completed on the adsorption of inorganic contaminants on single mineral phases, this work has primarily been undertaken using simple electrolyte solutions (e.g., $NaNO_3$ or $NaCl$) that are not representative of natural groundwaters. Agencies should fund research that studies sorption and the identification of surface species (by surface spectroscopy) for systems where the solutions contain the major ions found in natural groundwater (e.g., calcium, magnesium, carbonate, phosphate, silicate, and dissolved organic carbon). Adsorption constants for several major components are currently missing from the existing databases (or based on little experimental data), and are urgently needed to quantify the effects of competitive adsorption on contaminant surface reactions. In addition to competitive effects, the effects of major ion adsorption on the electrical double layer properties of minerals are still poorly understood. Although some data have been collected on competitive adsorption processes, there are data gaps and the capability to model such data with a self-consistent, comprehensive sorption model is still in question.

3.2.3.4 Application of Surface Complexation Modeling in Fractured Rock Systems

Modeling and laboratory studies of sorption in fractured rock systems need to be funded for improvement of reactive-transport simulations in this type of system. Although there have been various experimental studies of contaminant migration in fractured rock, comprehensive studies in which surface complexation model are used to describe contaminant retardation are scarce.

3.2.3.5 Application of Surface Complexation Modeling within the Vadose Zone

Modeling and laboratory studies of sorption in unsaturated zone systems need to be funded for improvement of reactive-transport modeling in vadose zone applications. The panel was not aware of a comprehensive study in which a surface complexation model was used to describe contaminant retardation in an unsaturated system.

3.2.4 Other Related Discussion Topics

A common concern among regulators and licensees is that characterizing a field site sufficiently to use a conceptual model for sorption other than the constant-K_d model is too costly or too time-consuming to develop. However, it needs to be understood that, if sorption strongly influences the transport of the contaminant to a point of compliance, the constant-K_d model cannot be scientifically justified if there are even small variations of key chemical parameters in time or space. The panel was of the view that a protocol could be determined to evaluate the complexity of field sites by examination of a specific list of field measurements that are required for development of a predictive multi-parameter adsorption model. The panel was also of the view that the costs and development time for such a geochemical evaluation would generally not be prohibitive. Some model parameters are not important at every field site. Some parameters may not be field-site specific, but rather taken from standardized databases. The field research suggested above (recommendation #1) should provide the necessary experience and guidance to identify the list of parameters that are required to be measured at each field site.

In laboratory research, sorption processes have usually been studied in simple systems (i.e., one mineral, one contaminant in simple aqueous solutions) and models have been developed to describe these systems with "unconditional" surface complexation constants. One approach for the development of a surface complexation model for field sites has been to conduct mineralogical studies at the site, and then use unconditional surface complexation binding constants from a standardized database. Standardized databases for the thermodynamic data for solubilities and aqueous complex formation are available (e.g., the Nuclear Energy Agency of OECD, Paris), and there are databases of surface complexation model parameters (e.g., Dzombak and Morel book on *Surface Complexation Modeling: Hydrous Ferric Oxides* and the online database of the Forschungszentrum Rossendorf, Germany). The simple systems studied thus far, however, are sufficiently different from soil-groundwater systems that the model parameters needed for a complete thermodynamic description are not available (e.g., carbonate and silicate surface species and associated electrical double layer parameters). Thus, there is still considerable work to be done to establish and promote the use of unconditional surface complexation binding constants to model field data in an internally consistent manner (see recommendation #3 above). Spectroscopic data (e.g., XAFS) are needed to characterize surface species that form on different minerals in variable aqueous solution compositions, especially solutions representative of groundwaters, and to determine which substrates adsorb a contaminant preferentially in natural soils.

An alternative approach is to develop knowledge and experience from the study of several field sites, extracting information regarding the most important processes involved in contaminant retardation, and develop simpler conceptual models that describe these observations (e.g., the USGS work at Cape Cod, Massachusetts and Naturita, Colorado). A key topic here is to develop an understanding of how site heterogeneity in porous medium properties and the spatial variability of chemical conditions in groundwater each affects K_d values in field systems.

Most modeling applications of contaminant migration at field sites assume that the local equilibrium assumption is valid for sorption processes. However, adsorption kinetics may need to be considered in systems with rapid fluid flow or in which mass transfer limitations control the approach to equilibrium. Hysteresis between adsorption and desorption processes has been documented in the laboratory; desorption is frequently slower than adsorption, and may be the most important process controlling groundwater contaminant concentrations in older field sites. The timescale of contaminant interactions with soil in the field is longer than that in the laboratory, and it is important to have developed techniques for site characterization that separate contaminants present in sorbed and precipitated states (see recommendation #1).

3.3 Nucleation, Precipitation, and Dissolution Processes

John A. Apps, Alex J. Blum, and Kenneth M. Krupka

3.3.1 Motivation
In many subsurface situations of consequence (e.g., contaminant hydrogeology, petroleum extraction, carbon sequestration, etc.), scientists and engineers are being asked by decision-makers to predict the fate of components of concern at the field scale under conditions where both reactions **and** transport are processes of first-order importance. These situations and the modeling tools and approaches that are used to analyze them were the subject of a workshop, "Conceptual Model Development for Subsurface Reactive Transport Modeling of Inorganic Contaminants, Radionuclides, and Nutrients," held at the La Posada in Albuquerque, New Mexico on April 20-22, 2004.

A major objective of the workshop was to examine and understand the "conceptual gap" between current modeling activities and conditions actually present in the field, and make recommendations as to how the gap could be closed through targeted research initiatives. In order to define research priorities for this field, the members of each breakout session were asked to evaluate the state of the art in order to define: a) advances needed in scientific understanding, b) new approaches for conceptual model development, c) improved methods for assessing the values of field-relevant reaction parameters, and d) suggestions for targeted research initiatives that would lead to new model development. Discussions from the breakout sessions on "Nucleation, Precipitation, and Dissolution Processes" are summarized here.

3.3.2 Background
Despite the importance of nucleation, precipitation, and dissolution (NPD) processes in realistically simulating both natural and anthropogenically induced geochemical processes in the field, they are infrequently incorporated in the current reactive transport models (RTM) in other than somewhat superficial or overly simplistic (e.g., local equilibrium) ways. Participants in the breakout session on NPD reactions represented a broad range of research expertise that included fundamental molecular-scale processes, principles of thermodynamics and kinetics, microbiology, field-scale geochemistry and computational modeling. They were asked to consider applications where model simulations of non-equilibrium NPD reactions in the environment would be important, the state of fundamental knowledge of the theoretical and empirical rate laws for describing these reactions, whether the fundamental knowledge is being effectively translated into computational models, and suggestions for targeted initiatives that would lead to improved representation of NPD processes in RTM's.

3.3.3 Applications Critically Dependent on NPD Processes
It is well recognized that NPD processes play a significant role in the fate and transport of chemical species in nature. They determine the mobility of chemical components, their interaction between other components in solution, and whether they become biologically active as either nutrients or toxins. Solid phases (minerals) act as reservoirs for chemical components, retarding the movement, and record a history of the physical, chemical and biological states of a system. NPD processes also alter the spatial distribution between the solid and fluid phases, which may modify critical physical and hydrologic properties of the system, such as porosity and permeability.

Members of the breakout session summarized a number of practical applications where knowledge of NPD processes is essential. These applications served as a starting point for identifying important knowledge gaps that must be addressed in order to develop better, and more practical RTM's for predicting system behavior. Potential applications in which NPD processes are critical include the following:

- Predicting the effects of agricultural land use practices on soil and water quality:
  - o Dissolution, availability and off-site transport of fertilizer phosphates and nitrates
  - o Understanding leaching and diagenetic changes in soils as a result of land-use practices
  - o Understanding diagenetic changes in soils due to wetting and drying cycles
- Leaching of radionuclides from soils, glasses and industrial by-products
- Subsurface injection disposal:
  - o Industrial organic and inorganic wastes (particularly acidic, basic, and metal-laden solutions)
  - o Waste brine streams from water treatment and oil recovery
  - o Sour gases (CO_2, SO_2, H_2S) and the performance of wells used for carbon sequestration
- Evaluating and improving the performance of reactive (redox, native iron) barriers for confining or attenuating migrating wastes
- Technologies for improving the extraction of oil, and soluble minerals through solution mining
- Water supply well and aquifer performance, and changes in water quality during injection and recovery of surface waters stored in subsurface aquifers
- Reactive transport and neutralization of acidic mine drainage
- Utilizing microbial processes to precipitate or dissolve solid phases in order to sequester subsurface contaminants
- Precipitation of salts by concentrating brines through evaporation
- Understanding natural, large-scale geochemical changes in the earth's subsurface

3.3.4 Advances Needed in Scientific Understanding

Although computational strategies can be developed to simulate many of the processes participating in the applications listed above, several significant, perhaps insurmountable, challenges remain. For example, quantifying rate constants for slowly reacting multi-component geochemical processes, taking decades to millennia to achieve equilibrium are significantly more complex than those of simple chemical reactions in which only a few stable phases participate, and where equilibrium is achieved within a few days. This is particularly true in systems where multiple metastable phase assemblages are possible and where it is not clear when and how equilibrium conditions are approached. In addition, important functions, such as stoichiometric relationships and rate expressions, are difficult to determine and can be subject to many experimental errors. These topics are subjects of ongoing research and significant advances are being made. However, the magnitude of the information needed to describe (parameterize) complex natural systems is very large, especially when considering the limited time available to address important environmental issues. Our current understanding of mineral dissolution and precipitation kinetics also suggests that there are no universal rate laws that can be integrated into reactive transport codes. Current concepts relating to NPD kinetics in RTM's must therefore be adapted to facilitate model testing and permit progress in understanding phenomena in the field.

Abundant thermodynamic data are available for pure end-member phases, common solid solutions and co-precipitates. These data provide a substantial basis on which to predict the equilibrium state of reactive systems. Despite the abundance of data, numerous gaps and inconsistencies remain. For example, the chemistry and thermodynamics of many secondary phases in natural systems, particularly clays, are difficult to quantify and describe. Furthermore, the accuracy of thermodynamic data is commonly insufficient for rigorous modeling, and data compilations are sometimes not internally consistent. These limitations must be rectified before kinetic schemes can be fully integrated in RTM's.

There is a relatively well-developed theoretical basis for the dissolution and precipitation kinetics of pure end-member phases. The dissolution rates of many minerals in dilute solutions have also been measured as a function of solution pH. Yet dissolution and precipitation processes, which are critical to natural systems, remain poorly understood. Many reactions occurring in natural systems are kinetically controlled and metastable phases are commonly dominant over more stable phases. Some of the issues requiring resolution are as follows:

- It has become apparent that many silicate reactions involve different mechanisms depending on the extent of approach to equilibrium, and whether equilibrium is being approached from supersaturation or undersaturation. Defect structures in crystals can play an important role in the kinetics of dissolution, which is commonly neither recognized, nor understood. The kinetics of reactions occurring near equilibrium is also poorly quantified, especially in relation to either congruent or incongruent dissolution.
- Metastable and poorly crystalline phases precipitating from supersaturation have a predominant role in determining the evolution of natural systems. Moreover, the rate laws for such phases have for the most part not been elucidated. In particular, the role of nucleation and the surface free energy contribution in the formation of metastable phases is not well understood. These processes provide a theoretical basis for the Ostwald Rule of Stages.
- The progressive transformation of metastable phases to more stable intermediate phases also exerts an important influence on the reactivity of systems, but is difficult to predict.
- The effects of both organic and inorganic trace constituents in retarding or accelerating crystal growth (including nanoparticles) in multi-component systems are often critical, but are poorly explored and poorly understood.

In addition to issues relating to precipitation and dissolution, there are several processes of considerable importance in many natural and anthropogenically modified systems where interfacial free energies exert a dominant control. Both the theory and thermodynamic parameters necessary to quantify these processes are poorly understood. They include:

- The relative importance of homogenous versus heterogeneous nucleation kinetics and mechanisms
- The stabilization of nanoparticles (e.g., colloids as a function of solution composition and pH)
- The kinetics of mineral coarsening (i.e., Ostwald ripening) and aggregation
- Biomineralization and biologically mediated dissolution and precipitation

Finally, there are physical processes, which may exert important controls on system hydrology, but are not well characterized. Examples include:

- Hydrologic routing in the unsaturated zone
- The effects of wetting/drying cycles in the soil zone on dissolution, precipitation, and colloidal transport
- Changes in porosity and permeability of soils and sediments, because these properties are

often directly linked to location and rate of dissolution and precipitation reactions

3.3.5 New Approaches for Conceptual Model Development

The dependence of reaction kinetics on a large number of specific and detailed chemical and physical conditions makes it difficult to identify any set of universally applicable rate laws, and it can be difficult even to identify the predominant kinetic controls in specific systems. Many of these processes are also difficult to study experimentally, largely because of the long time scales of many kinetically controlled reactions. It was therefore concluded that conceptual model development must go hand-in-hand with the refinement of RTM codes so that reaction kinetics can be integrated while concurrently identifying needed refinements in RTM's. Such parallel development is a necessary requirement if natural and artificial phenomena in the field are to be interpreted.

RTM codes using different techniques and designed for different applications are currently available. The results of theoretical and experimental research on dissolution and precipitation rate laws could be more readily incorporated if codes were more versatile, more user friendly, and had better documentation. Testing different kinetic rate laws in RTM codes would favor those that are easier to modify and manipulate. In particular, members believed that codes should be modularized so that different kinetic/stoichiometric models could be inserted and tested at will. To enable this conceptual testing, it would be helpful if a compilation was available describing and explaining the principal features of the various extant codes, so that knowledgeable investigators could more easily test and evaluate the appropriateness of specific kinetic models. Finally, it was concluded that RTM codes are needed in which NDP processes are coupled with hydrologic effects such as changes in porosity and permeability, and coupling between fluid flow and chemical processes.

Another aspect aiding conceptual model development would be the availability of internally consistent databases. These databases should be expanded beyond traditional thermodynamic compilations for solids and aqueous species to include data relating to precipitation and dissolution kinetics, interfacial free energies and affinity terms. It was recognized that substantial challenges would attend such endeavors, yet the availability of such databases would greatly aid in the development of conceptual models. Such databases should be well documented, be internally consistent, and transferable between codes. Additional databases could be established for parameters derived through model-based interpretations, especially those relating to hydrologic phenomena impacted by NPD processes.

It was also believed that specific field sites should be chosen for further testing. Members singled out for emphasis, sites permitting the evaluation of phenomena in partially saturated zones and the evaluation of aging sequences in soils. Conceptual model development would be enhanced through the selection of a site or sites that would allow for modeling and characterization of experiments at multiple spatial and temporal scales, ranging from fractions of a meter to kilometers, and over time domains from short controllable periods (i.e., days to a few years) to passive observations of phenomena operating over time scales extending to millennia and beyond.

3.3.6 Improved Methods for Assessing Field Relevant Parameters

The importance of feedback between theoretical developments, experimental studies and field observations is widely recognized. In some cases this feedback involves using inverse techniques to determine the effective magnitudes of parameters that are not otherwise accessible using independent experimental methods. There was a general consensus among members that detailed field studies are presently a critical research priority in identifying and quantifying dissolution and precipitation mechanisms and kinetics in natural environments. Because of the potential multitude of important parameters in reaction kinetics, more field observations to constrain both the overall rates and the mechanisms of mineral reactions in natural systems and correlate these findings with supporting experimental data would greatly help focus both experimental and theoretical studies. Observational information acquired from the field should also be organized to show correlations between existing solid phases or solid phases in transition, and the physical, chemical and biological regimes under which they exist. Such field studies would be enhanced by:

1. Better and more standardized sampling techniques to preserve field characteristics and avoid artifacts
2. Characterizing changes in solution chemistry and mineral abundances along hydrologic transects
3. Attention to the presence of sub-micrometer sized particles (i.e., colloids or nano-particles) and their impact on macroscopic equilibrium or steady state solution chemistry
4. More detailed characterization of changes in solution and mineral chemistry and abundance
5. Characterization of fluids within unsaturated environments
6. More measurements to obtain precipitation/dissolution rates in the field
7. Intermediate-scale characterization of localized dissolution and precipitation of solids (e.g., worm holing, fingering, stability of reaction fronts)
8. Localized changes in porosity and permeability due to differential dissolution and precipitation mechanisms
9. Better measurements of relative states of equilibrium with respect to the mineral phases present in soils, sediments and bedrock
10. General characterization of mineral alteration (depth profiles of elements in co-precipitates, the composition and degree of order in amorphous coatings, and their correlation with coexisting solution chemistry)

With respect to supporting laboratory studies, it was concluded that it would be more productive for experimental work to focus on testing systems undergoing rapid change, as analogues for natural systems where the kinetics of transformation are slower. The time and resources needed to study slowly reacting natural systems should await a better research focus from more detailed field observations. There is a strong coupling between physical processes affecting mass transport and the fundamental thermodynamic and kinetic laws at solid solution interfaces. Intermediate-scale experiments with systems that simulate the dominant precipitation/dissolution reactions could play an important role in helping to develop and test precipitation/dissolution components of RTM's.

3.3.7 Targeted Research for New Model Development
The NPD issues reviewed by participants in the breakout session lead to a series of suggestions for research activities necessary for new model development. These activities are summarized below.

1. Develop algorithms describing different mechanisms controlling mineral dissolution and precipitation in a form suitable for incorporation in a reactive transport code
2. Develop algorithms describing the kinetics of processes that lead to the stabilization of metastable phases in systems under supersaturated conditions (i.e., algorithms that provide a theoretical kinetic basis for the Ostwald rule of stages)
3. Develop algorithms that couple the changes in permeability, and subsequent changes in the

transport of precipitation/dissolution reaction components, that are induced by precipitation, dissolution, and particle filtration events

4. Identify and evaluate methods for acquiring data on interfacial free energies for mineral families where multiple metastable phases have been observed in the natural and anthropogenically modified environments (e.g., carbonates, sulfates, phosphates, silicates, and sulfides)

5. Develop algorithms that reflect changes in the interfacial free energies of minerals as a function of the coexisting aqueous solution properties

6. Incorporate algorithms in models that simulate the development of particle size distributions

7. Explore the impact of foreign molecular and ionic species adsorption on mineral surfaces, and how this may affect dissolution and precipitation kinetics; develop a theoretical basis, and identify needed chemical parameters to quantify the dependence of kinetics on adsorbed species

8. Incorporate the algorithms suggested above in a suitable RTM code for conceptual testing and evaluation

9. Test models simulating metastable phase formation against hypothetical examples and field sites where such phenomena are evident

10. Develop and test models reflecting nucleation phenomena, and compare predicted behavior against corresponding field conditions where such phenomena are operative

3.4 Redox Processes

Gary P. Curtis and John M. Zachara

3.4.1 Motivation
In many subsurface situations of consequence (e.g., contaminant hydrogeology, petroleum extraction, carbon sequestration, etc.), scientists and engineers are being asked by decision-makers to predict the fate of components of concern at the field scale under conditions where both reactions and transport are processes of first-order importance. These situations and the modeling tools and approaches that are used to analyze them were the subject of a workshop, "Conceptual Model Development for Subsurface Reactive Transport Modeling of Inorganic Contaminants, Radionuclides, and Nutrients," held at the La Posada in Albuquerque, New Mexico on April 20-22, 2004.

A major objective of the workshop was to examine and understand the "conceptual gap" between current modeling activities and conditions actually present in the field, and make recommendations as to how the gap could be closed through targeted research initiatives. In order to define research priorities for this field, the members of each breakout session were asked to evaluate the state of the art in order to define: a) advances needed in scientific understanding, b) new approaches for conceptual model development, c) improved methods for assessing the values of field-relevant reaction parameters, and d) suggestions for targeted research initiatives that would lead to new model development. Discussions from the breakout sessions on "Redox Processes" are summarized here.

3.4.2 Background
Reactive transport models are useful tools that can be used to integrate diverse sources of information into a scientific framework that can then be used to test hypotheses about the processes that control the past and future behavior of a groundwater environment. A critical aspect of reactive transport modeling is identifying and quantifying the key chemical species and their associated reactivity in a given environment, and the coupling of this information with fluid flow. This fundamental requirement of defining the reactants and reactivity is a unifying concept that applies to several classes of reactions including precipitation-dissolution, adsorption and ion exchange, and redox reactions. The breakout group on redox processes focused most of its discussion on identifying redox active species, describing the reactivity associated with an often diverse microbial population, and improving existing conceptual and numerical models of reactive transport involving redox active species and microorganisms.

Redox reactions are an important class of reactions that involve the change of valences of two or more reactants. The importance of redox reactions derives from the significant affect that the valence state of an element has on the chemical reactivity of that element including its toxicity and mobility in the environment. For example, Cr(VI) is used in many industrial processes and is a common groundwater contaminant. In oxidizing groundwater Cr(VI) is present as oxyanions (CrO_4^{2-}, $HCrO_4^-$, $Cr_2O_7^{2-}$) that are relatively mobile and can be toxic. However, if Cr(VI) is reduced to Cr(III) (a micronutrient) the ions in solution are predominantly cationic (Cr^{3+}, $Cr(OH)^{2+}$, $Cr(OH)_2^+$) and their total concentration is relatively small because insoluble precipitates form in most environments. This precipitation combined with the binding of the Cr(III) to surfaces and organic matter significantly reduces the mobility of Cr in the environment. Additional examples of elements subject to diverse changes in reactivity include transition metals (Co, Mn, Fe), metalloids (Se, As), actinides (U, Np, Pu), and several other elements (O, S, N, and V).

It was recognized in the group discussions that microorganisms play a central role in mediating most redox reactions in the subsurface and that the subsurface geochemistry and microbiology of a given environment are intricately coupled. For example, the geochemistry of an environment, in particular the presence of electron acceptors such as Fe(III) oxides and SO_4 (either dissolved or present in solid phases) can determine the type of microorganisms that will dominate the local environment. Conversely, microbial processes can alter local environments either directly or indirectly by causing (1) the formation of new species that can form new phases including solid precipitates and CO_2/CH_4 bubbles, (2) the reductive dissolution of Fe(III) and Mn(III/IV) oxide coatings which can be important adsorbents and redox/proton buffers, and (3) the change of solution pH, alkalinity, and ion composition and concentration. Each of these processes, coupled with the complicating effects of transport, suggest that the reactivity of many subsurface redox environments can change temporally and spatially.

Redox reactions and microbially mediated processes occur on scales that vary by many orders of magnitude (Figure 1). Redox reactions, by definition, occur on a molecular scale (10^{-10} to 10^{-9} m) often between a heme group (Fe-S redox center) in a microbiologic redox active protein and a dissolved, oxidized inorganic ion or an oxidized moiety in or on a mineral surface. Modeling reactions at this scale is generally accomplished using *ab initio* or molecular dynamics simulations. These approaches give important information and insight about mechanisms of redox reactions and factors governing their kinetics and energetics. However, it is infeasible to extend these simulation techniques to the laboratory or field scale because of the enormous number of atoms involved in the simulation, and the importance of other regulating environmental processes. Thus, as with many reactive processes in the environment, there is a challenge of how to scale reaction rates that necessarily occur on the molecular scale to scales on the order of 0.1 to 1 m. This latter scale is approximately the scale of redox zonation observed at many sites contaminated by hydrocarbons and other readily metabolizable organic substrates. Scaling the rates of microbially mediated reactions could require special considerations because as the scale increases, the enzyme(s) responsible for mediating redox reactions in single cells at the scale of 10^{-6} m may occur in diverse species in microcommunities, consortia, and biofilms at the scale of 10^{-4} to 10^{-3} m, and may be distributed in complex patterns in ecosystems at the largest scales. Ideally, experiments and simulations would be conducted over a wide range of scales and information at one scale would be transferable or consistent with information at other scales.

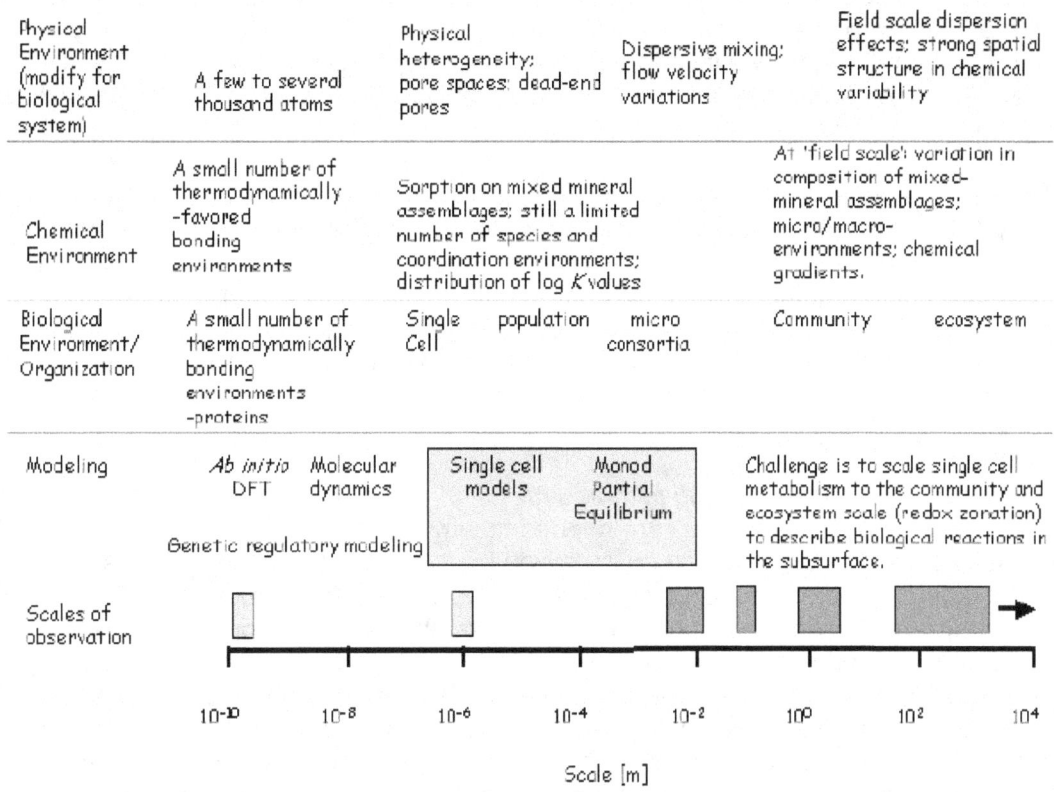

Physical Environment (modify for biological system)	A few to several thousand atoms	Physical heterogeneity; pore spaces: dead-end pores	Dispersive mixing; flow velocity variations	Field scale dispersion effects; strong spatial structure in chemical variability
Chemical Environment	A small number of thermodynamically-favored bonding environments	Sorption on mixed mineral assemblages; still a limited number of species and coordination environments; distribution of log K values		At 'field scale' variation in composition of mixed-mineral assemblages; micro/macro-environments; chemical gradients.
Biological Environment/ Organization	A small number of thermodynamically bonding environments -proteins	Single Cell population micro consortia		Community ecosystem
Modeling	*Ab initio* DFT Molecular dynamics Genetic regulatory modeling	Single cell models Monod Partial Equilibrium		Challenge is to scale single cell metabolism to the community and ecosystem scale (redox zonation) to describe biological reactions in the subsurface.
Scales of observation	10^{-10} 10^{-8}	10^{-6} 10^{-4}	10^{-2} 10^{0}	10^{2} 10^{4}

Scale [m]

Figure 1. Scale issues for biologically mediated reactions and transport.

3.4.3 Research Needs: Conceptual and Numerical Reactive Transport Models
3.4.3.1 Redox speciation
As previously discussed, one of the fundamental requirements for conducting reactive transport simulations is that all of the important chemical species must be identified and quantified. This knowledge is a prerequisite to identify the reaction stoichiometry that accounts for the formation and consumption of each of the species and provides the basis for conducting reactive transport modeling. Subsequent to identifying a particular new species, thermodynamic, and/or kinetic data must be obtained so that reactive processes can be incorporated into a quantitative reactive transport model. For example, thio-arsenic species were first postulated based on solubility calculations, then they were identified by spectroscopy, and now their formation constants are being established. Speciation calculations subsequently demonstrated that these thio-arsenic species are the dominant forms of arsenic under certain conditions. Application of new and existing spectroscopic techniques could identify new chemical species for other chemical elements. These discoveries are a critical first step for improving conceptual models for the transport of both organic and inorganic species in groundwater.

3.4.3.2 Geomicrobiological characterization (tools to link activity to reactive transport models)
Microbial growth and decay is either not included in current reactive transport models or they are based on the implicit assumption that microbial activity is proportional to the microbial populations (primarily represented as biomass). Although this proportionality has been observed in certain circumstances, many exceptions have also been observed. In addition,

several parameters used in simulating microbial growth and decay, specifically cell yields and death rate may vary significantly with time and environmental factors. These parameters, however, are often assumed to be invariant in reactive transport simulations. This overly simplified approach to quantifying reactivity in a microbial environment is a key factor limiting the applicability of reactive transport models.

New techniques are needed that directly probe the activity (or reactivity) of particular microbial population, particularly in relation to their function or to the geochemical reaction(s) that they mediate. These techniques could be based on work conducted at scales on the order of the size of a cell or smaller that focus on identifying reactive groups (enzymes) within a cell and the processes that regulate these enzymes. Development of tools that measure microbial activity *in situ* at these scales is needed. Characterization of the types and kinetic properties of these reactive groups could yield improved measures of microbial activity applicable at the field redox zonation scale.

3.4.3.3 Development of new modeling techniques
New conceptual and numerical models are needed to simulate microbially mediated reactions and these models need to be based on an improved understanding of (1) the factors that control the activity of different bacterial functional groups in a particular environment and (2) the explicit relationships that exist between microbial activity, biogeochemical reaction kinetics, and the chemical speciation of redox sensitive elements. Models that could be considered could vary significantly with the scale and mechanistic detail of the processes being modeled and modeling approaches. Ideally, results at one scale should provide insight for simulations conducted at other scales. The models to be considered could include gene regulation models, whole-cell metabolism and energy balance models, and microbial community models. The microbial community models are the most relevant to field scale reactive transport simulations. These models need to be able to accurately simulate consumption of multiple terminal electron acceptors without the use of empirical constants and simulate the selection and adaptation of selected physiological functions from key simulated environmental parameters. Moreover, the models would have to explicitly link electron acceptor consumption with nutrient availability/flux and the geochemical reactions that they directly and indirectly mediate. Because these models are usually highly complex, it is desirable that they be easily, and perhaps automatically, simplified depending on the geochemical conditions.

3.4.3.4 Field site and supporting facilities
Participants strongly endorsed the identification and development of a field site for conducting reactive transport experiments by the multi-agency research community that focuses on redox reactions. This site could provide meaningful samples for biogeochemical speciation analyses by advanced spectroscopic techniques and provide opportunities to assess the performance of new geomicrobiological characterization and modeling tools being developed. The field site needs to be well implemented, managed and supported with onsite laboratories and be large enough to permit both small-scale and large-scale biogeochemical experimental programs to be conducted.

This user facility will allow researchers to test new innovative characterization tools (e.g., Section 2) and new reactive transport models (e.g., Section 3) in a field environment. The field site will also allow users to conduct rigorous experiments to explore and illuminate fundamental processes that control redox reactions in the subsurface. The site can also be used to test innovative ideas for manipulating subsurface redox conditions for the purpose of remediating contaminated aquifers and sequestering contaminants. Ideally, two sites would be available;

one in porous medium and a second in fractured rock.

3.4.3.5 Demonstrate effective means of model parameterization

Numeric models for reactive transport involving microbially mediated redox reactions are often based upon a large set of reaction parameters. In current models, these parameters include Monod kinetic constants, cell yield and cell death rates and, perhaps, inhibition parameters which account for competition among different microbial populations for a given substrate or electron acceptor. There are also rate constants for the associated suite of geochemical reactions that are indirectly regulated by microbial activity (e.g., those resulting from biogenic changes in pH, CO_2(gaseous), ion composition, or solute valence). Because of the large set of reaction parameters, it can be difficult to calibrate a reactive transport model and even when a good calibration is obtained, it is possible that the calibration is not unique. New methods for calibrating site-specific models using information obtained at different scales to constrain semi-empirical reaction parameters would be helpful. These methods may be built upon global optimization methods, data fusion, sensitivity analyses, and possibly, using the theory of the design of experiments. A full demonstration of how model parameters can be determined either from independent experiments, multi-scale reactive transport modeling or from model calibration would advance the credibility of reactive transport models with the scientific, regulatory, and applied communities.

3.4.3.6 Evaluate simplified models

Reactive transport models of microbially mediated redox reactions are likely to be very complex and require estimates or measurements of many model parameters. Currently, some such models contain 20 to 40 parameters probably because of the complexity of the underlying processes. New models will likely have at least as many model parameters that can make the application of these models difficult. Methods need to be developed that allow the user to easily derive simple models from the more scientifically rigorous models. Simplification can make the model easier to apply, the results more transparent, and sensitivity analyses more straightforward. Simplified models must, however, retain scientific credibility, and meaningful guidelines should be established by the knowledgeable research community. The simplification of the conceptual model could occur at one of two levels. In the first instance, the user could derive a simpler model from a more complicated model. Alternatively, it is plausible that new models developed will be capable of adapting the numerical solution for systems dominated by redox reactions that evolve spatially and temporally during the reactive transport simulation.

3.4.4 Practical Needs: Application of Existing Models and Approaches

Several practical needs were identified that could facilitate the application of existing models to either research activities or remedial activities. These needs include additional documentation, algorithm development, new model calibration approaches, and demonstration of the utility of these modeling approaches. However, the scope and time requirements of this new work are likely to be sufficiently small so that a significant impact could be made in a relatively short time period.

3.4.4.1 Guidelines for practitioners

There are currently only a small number of applications of reactive transport simulations where redox reactions were a dominant process. These applications are primarily described in the scientific literature as research articles. It would be helpful if the basic findings from all of the previous applications of reactive transport model would be summarized in a guidance document prepared by several relevant federal agencies. This guidance document would summarize approaches for identifying the key reactive species and reactions, describe possible kinetic equations, provide ranges of parameter values, discuss common pitfalls of reactive transport modeling, and illustrate approaches for evaluating model sensitivity and prediction uncertainty. Attributes and contributions of the reactive transport modeling to scientific understanding and decision-making would also be highlighted.

3.4.4.2 Enhanced models
Access to more flexible, sophisticated and user-friendly models would facilitate application of the reactive transport modeling of redox reactions to more field sites. Such models would allow the user to more readily test alternative hypotheses so that greater insight can be gained in a shorter period of time.

3.4.4.3 Access to realistic ranges of parameters
One of the challenges of conducting reactive transport modeling of microbially mediated reactions is estimating each of the large number of parameters required by simulations. In many instances, it may be desirable to select a parameter from the literature rather than attempt to estimate that parameter from model calibration against appropriate field or laboratory data. The importance of these parameters would need to be evaluated by a sensitivity analysis. Having access to realistic ranges of parameters relevant to field conditions being considered (e.g., oxygen concentrations, water composition, mineralogy, temperature, linear groundwater velocities, etc.) would be useful in circumstances where calibration data are limited or where parameter values are highly covariant.

3.4.4.4 Consideration of conceptual model and parameter uncertainty
Approaches for comprehensively assessing the uncertainties involved in reactive transport modeling resulting from parameter uncertainty and from incomplete site characterization need to be summarized. Approaches used in problems of groundwater flow and nonreactive transport could be adapted to the problems of reactive transport and subsurface biologic community evolution as it relates to redox status and reactivity. However, many of these approaches are based on Monte Carlo techniques which may not be feasible for biogeochemical reactive transport simulations because of the excessive computational burden. The meaning and implications of the uncertainties obtained by these techniques must be presented in a clear and accessible manner to scientists, engineers, regulators, and decision makers.

4. PROCEEDINGS AND ABSTRACTS

This section of the report includes the proceedings and extended abstracts for the formal presentations from the reactive transport workshop. The papers and abstracts are presented in the order of oral presentation based on the workshop agenda (see Section 2.). Electronic copies can be obtained from the public web pages of the Interagency Steering Committee on Multimedia Environmental Models (www.iscmem.org) through the link under the Conference Proceedings heading. Additionally, when made available by the author, electronic PDF versions of the workshop presentations can be examined through the same web link.

4.1 Geochemistry, Groundwater and Pollution: "Learning by Modeling"

C.A.J. Appelo[1], A. Dimier[2]

[1]*Hydrochemical Consultant, Amsterdam, Netherlands, appt@xs4all.nl*

[2]*Andra, Chatenay Malabry, France*

1 INTRODUCTION

Predicting the fate of chemicals that leach from waste deposits and deciphering their transport in the invisible subsoil must rely on 1) understanding the pattern of groundwater flowlines and travel times, 2) insight in the chemical reactions which retard the solutes, and 3) the quality of the database of the geochemical model. Inhomogeneities in an aquifer can be deduced from tritium/helium age profiles with depth, and otherwise from costly tracer experiments. The retardation of heavy metals is illustrated for a case of acid mine drainage from a former uranium mine. In addition to retardation by sorption, U(6) may disperse into confining layers where it reduces to U(3) and precipitates. Lastly, a model is presented for the release of arsenic in Bangladesh groundwater. With a recently defined carbonate surface complex in the database, carbonate appears important for displacing sorbed arsenic to poisonous aqueous concentrations.

2 TRANSPORT, TRITIUM/HELIUM AGE DATING UNRAVELS IRREGULARITIES

The flowlines in a phreatic aquifer may follow the pattern shown in Figure 1. In the homogeneous case, the age of water varies with depth according to:

$$t = \frac{-D\,\varepsilon_w}{P}\ln\left(\frac{d}{D} - 1\right)$$

where t is time (yr), D is thickness of the aquifer (m), ε_w is water-filled porosity (-), P is recharge (m/yr), and d is depth (m) (Appelo and Postma, 2004). However, the ages calculated from radioactive decay of ^3H to its daughter ^3He in borehole DFG1 deviate from the ideal line for a homogeneous aquifer (fig. 1). The water is too old in the upper part and too young in the lower part, and seems mixed over depth. Clearly, the aquifer is not homogeneous, although the sediment samples indicate a uniform sand. We can, with a few simple formulas, derive the flow properties of the inhomogeneity that may explain the

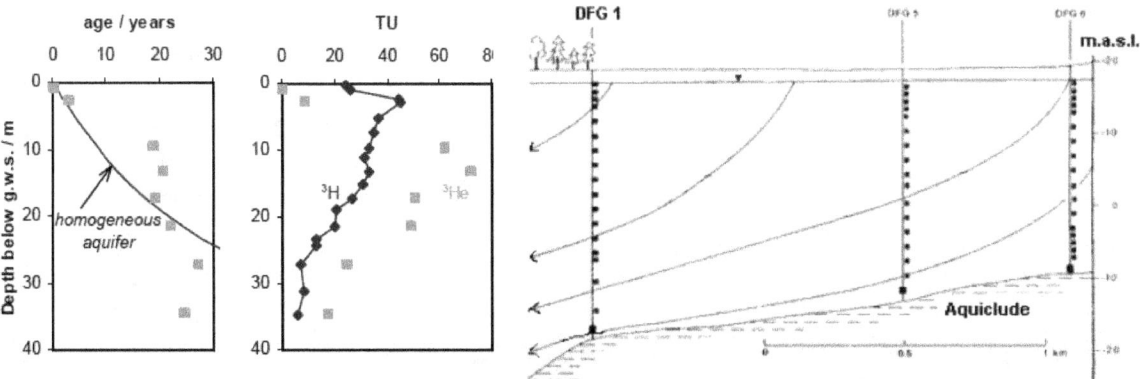

Figure 1. Regular flowlines in a phreatic aquifer (right) are contradicted by the age profile calculated from tritium/helium measurements in borehole DFG 1 (modified from Leuchs, 1988).

observed ages. For example, mixing over half of the depth in borehole DFG1 (about 18 m) requires a flowpath of $L = (18/2)^2 / 2\alpha = 810$ m, for a (vertical) transversal dispersivity $\alpha = 0.05$ m. We can also estimate that the zone upstream, where mixing takes place, has a permeability that is $(D - z) / z$ larger than of the average sediment, z being the thickness of the permeable layer. These numbers are useful for initiating a hydrogeological model in which irregularities are tried out and distributed over the section to simulate the observed age pattern. However, an exact and detailed flowpattern requires a much denser net of boreholes.

3 RETARDATION OF U(6) AND ZN FROM ACID MINE DRAINAGE

After closure of the uranium mine at Königstein (G), the acid mine drainage will enter the aquifer above the mine and flow towards the Elbe river (Bain et al., 2001). We like to predict how quickly the heavy metals and uranium advance and whether they are retarded with respect to water flow, $v_i = v_{H_2O} / R$. Traditionally, retardation is calculated from $R = 1 + q/c$, where q is sorbed and c is solute concentration (mol/L). However, for transport the flushing factors given by $(\partial q_i / \partial c_j)_{i \neq j}$ should be considered (the slope of the isotherm if $j = i$), rather than the distribution coefficient $K_d = q/c$.

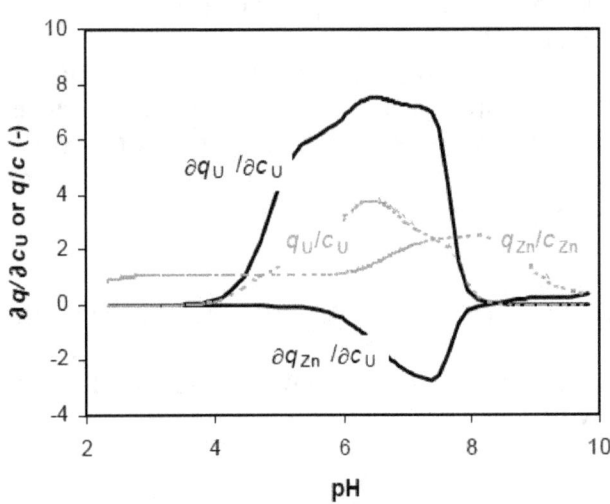

Figure 2. Distribution coefficients (q_i/c_i, dotted lines) and flushing factors ($(\partial q_i / \partial c_U)_{i \neq U}$, full lines) for Zn and U(6) in acid mine drainage as function of pH.

In the Königstein case the distribution coefficients differ markedly from the flushing factors, shown for Zn and U(6) as function of pH in Figure 2. In the calculations, the mine water composition of Bain et al. was used and the WATEQ4F database of PHREEQC (Parkhurst and Appelo, 1999), with the exchange and surface complexation capacity set to $X^- = 50$ mM, Hfo_w = 2 mM and Hfo_s = 0.05 mM.

The distribution coefficient of Zn is determined by ion exchange at low pH, increases at intermediate pH's by sorption on iron-oxyhydroxides, and dwindles again at high pH through aqueous complexation. The distribution coefficient of U(6) shows the same pattern but is zero at low pH since the element is not exchanged (at least not in the WATEQ4F database). The slope $(\partial q_U / \partial c_U)_{i \neq U}$ is higher than K_d, indicating a concave isotherm for U(6) (at a fixed pH). Consequently, the U(6)

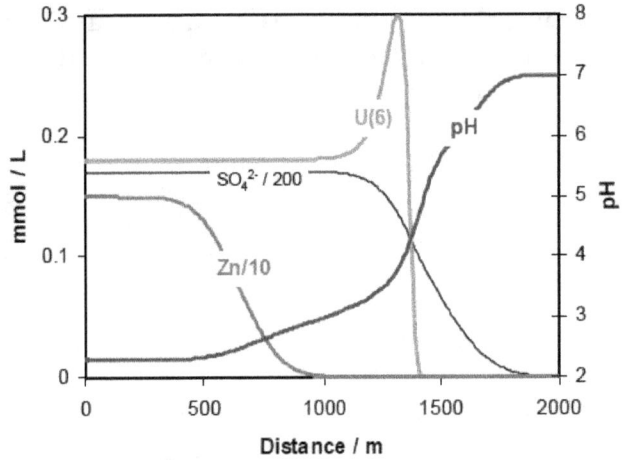

Figure 3. Concentrations of SO_4^{2-}, Zn^{2+}, U(6) and pH from acid mine drainage along a flowline calculated with PHREEQC-2.

front will be sharpening when concentrations decrease. On the other hand, for Zn the slope $(\partial q_{Zn} / \partial c_U)_{i \neq U}$ is negative in the pH range where U competes with Zn for sorption sites with an even stronger effect than of Zn itself.

The calculated concentrations along a flowline (fig. 3) show almost conservative behavior for SO_4^{2-} and a retardation of 2.1 for Zn by ion exchange (the retardation is not affected by U in the low pH range, cf. Figure 2). The concentration of U shows an increase as if a heap of snow is pushed up in front of a snowplough. The typical shape is related to the pH variation at the front. Uranium is sorbed at neutral pH, but as the pH decreases to below 4 in the acid mine water, U is desorbed again, and the decreasing concentration gives a sharp front.

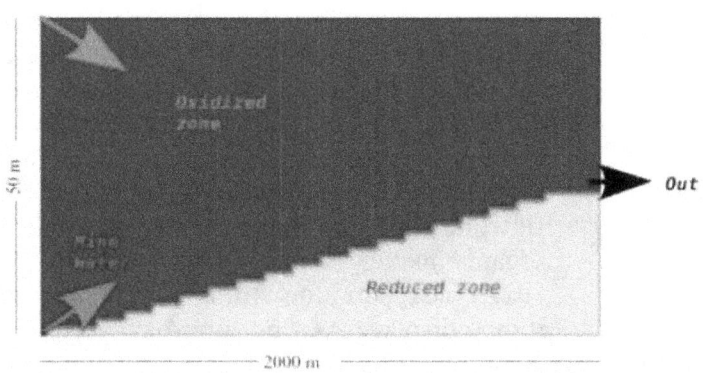

Figure 4. Outline of the model aquifer. The reduced zone contains pyrite.

A completely different picture is predicted for the aquifer, where the confining layers contain pyrite that reduces U(6) to U(3). The latter forms insoluble compounds which generates an infinite retardation for U. A rectangular cross-section of the aquifer, 50 * 2000 m above the Königstein mine was discretized in 40 * 100 cells for modeling with MT3D/PHREEQC (fig. 4). Acid mine water enters the section from the lower left corner and is accompanied by an equal flow of natural water from the upper left corner. The water leaves the section in the middle right. The lower right quarter, from the mine until the outflow cell contains pyrite.

The solute concentrations of U indicate the loss towards the reduced zone where coffinite precipitates, while at the front the snowplough effect appears (fig. 5). The contact with pyrite is determined by the vertical dispersivity, here taken to be 1 m. Probably, the translation towards the reduced zone enriched the strata with uranium in the geological past.

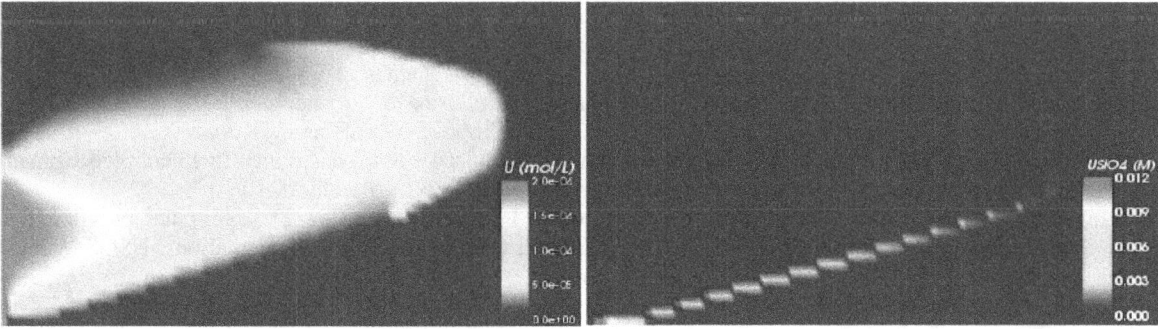

Figure 5. Modeled concentrations of U(6) in acid mine drainage entering an aquifer (left) and precipitation of USiO$_4$ (coffinite, right) at the border of the reduced zone. Calculated with Alliances, an MT3D/PHREEQC combination developed at Andra.

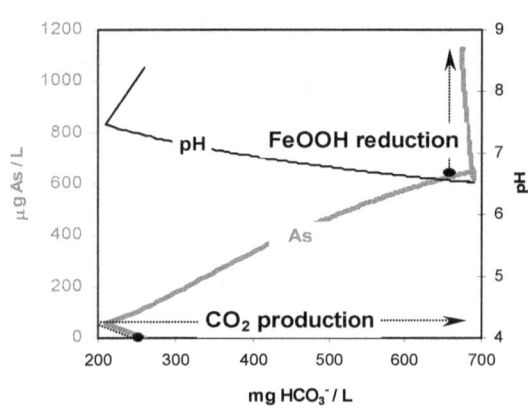

Figure 6. Increasing HCO_3^- concentrations will desorb As from iron-oxyhydroxide (Appelo et al., 2002).

The model results crucially depend on the validity of the chemical reactions and associated constants in the database. The surface complex of HCO_3^- has been avoided by experimenters and was neglected in databases, but it is rather essential for modeling sorption variations of heavy metals in nature since it is the dominant anion in groundwater and competes with other solutes for surface sites. The release of As in Bangladesh aquifers has been explained by the reductive dissolution of iron-oxyhydroxide (Kinniburgh and Smedley, 2001). However, if we consider a flowline that infiltrates in a soil where organic matter is oxidized aerobically, the HCO_3^- concentration strongly increases and it may displace As from iron-oxyhydroxides which have sorbed the element from surface water, but at a much smaller HCO_3^- concentration (fig. 6). The figure shows that the increase of HCO_3^- by CO_2 production in the soil can be accompanied by increase of As to 600 ppb, far above the drinking water limit (cf. details in Appelo et al., 2002). In this case, the effect is only notable in the model calculations if the surface complexes of carbonate are included in the database.

REFERENCES

Alliances, couplage chimie et transport avec MT3D/PHREEQC. Under development at Andra, France. Contact Alain.Dimier@Andra.fr.

Appelo, C.A.J. and Postma, D., 2004. Geochemistry, groundwater and pollution, 2nd ed. in prep.

Appelo, C.A.J., Van der Weiden, M.J.J., Tournassat, C. and Charlet, L., 2002. Surface complexation of ferrous iron and carbonate on ferrihydrite and the mobilization of arsenic. Environ. Sci. Technol. 36, 3096-3103.

Bain, J.G., Mayer, K.U., Blowes, D.W., Frind, E.O., Molson, J.W.H., Kahnt, R. and Jenk, U., 2001. Modelling the closure-related geochemical evolution of groundwater at a former uranium mine, J. Contam. Hydrol., 52, 109-135.

Kinniburgh, D.G. and Smedley, P.L. (Eds.), 2001. Arsenic contamination of groundwater in Bangladesh. British Geological Survey (Technical Report, WC/00/19. 4 Volumes). British Geological Survey, Keyworth.

Leuchs, W., 1988, Vorkommen, Abfolge und Auswirkungen anoxischer redoxreaktionen in einem pleistozanen Porengrundwasserleiter. Bes. Mitt. Deutsche Gewass. Jahrbuch Nr 52, Düsseldorf, 106 p.

Parkhurst, D.L. and Appelo, C.A.J., 1999. User's guide to PHREEQC (version 2)-a computer program for speciation, batch-reaction, one-dimesnional transport, and inverse geochemical calculations. USGS Water-resour. Inv. Rep. 99-4259.

4.2 Reactive-Transport Prediction Uncertainty and Simulation Accuracy, Observation Errors, and Sensitivity Analysis

Mary C. Hill

U.S. Geological Survey, Boulder, Colorado, USA, mchill@usgs.gov

1 INTRODUCTION

Prediction uncertainty is the likely discrepancy between model predictions and the actual, unrealized system responses. Contributions to uncertainty include solution error and limited capabilities of numerical models, error and deficiency of data, and errors in system conceptual models. Uncertainty can be reduced by improving numerical models and using numerical models, data, and conceptual models together. For example, using conceptual models to build simulations forces ideas about system behavior that are often vague and possibly wrong to be clarified and tested thoroughly against data. Problems with the numerical methods or constitutive relations, however, can obscure results. This abstract present three issues important to this process: (1) a common numerical-methods issue for ground-water transport simulations; (2) the problem of matching data too closely; and (3) using a model to evaluate the importance of observations to parameters, parameters to predictions, and observations to predictions.

2 A COMMON NUMERICAL-METHODS ISSUE FOR GROUND-WATER TRANSPORT SIMULATIONS

A common numerical issue in transport models is numerical dispersion. Mehl and Hill (2001) investigated the effects of numerical dispersion in the simulation of conservative transport on parameter estimation. The investigation used results from a two-dimensional laboratory experiment constructed of discrete, randomly distributed, homogeneous blocks of five sands. Measured hydraulic conductivities varied over more than two orders of magnitude; measured dispersivities varied over more than one order of magnitude. The five dispersivity values were not estimated due to insensitivity. The small amounts of numerical dispersion evident in Figure 1A resulted in significantly different optimized values of hydraulic conductivity and the different breakthrough curves shown in Figure 1B. Slightly better fits were achieved for the methods with more numerical dispersion, suggesting that the measured dispersivities are consistently too small. Basically, the estimated hydraulic conductivities are making up for the bias in the measured dispersivities, and methods with larger numerical dispersion require less adaptation. If the measured dispersivities were more accurate, the methods with less numerical dispersion would produce the more accurate results. In general, the bias is unknown, and it is advantageous to estimate dispersivity. In Mehl and Hill (2001), the insensitivity is addressed by lumping the five dispersivities and estimating a single value.

3 THE PROBLEM OF MATCHING DATA TOO CLOSELY

Closer correspondence between the simulation and measurements often indicates the model more accurately represents a system. However, when models are calibrated, predictive capability can be degraded by fitting measurements too closely, as shown in Figure **2**. This can occur when the model is overparameterized and close model fit is achieved by fitting the errors in the data. Thorough evaluation of data errors and the possibility of overfitting are critical. This is especially true for methods in which many parameters are defined. In these situations generally overfitting can be controlled using prior information and smoothness constraints, but the consequences of these methods may not be well understood by the modeler.

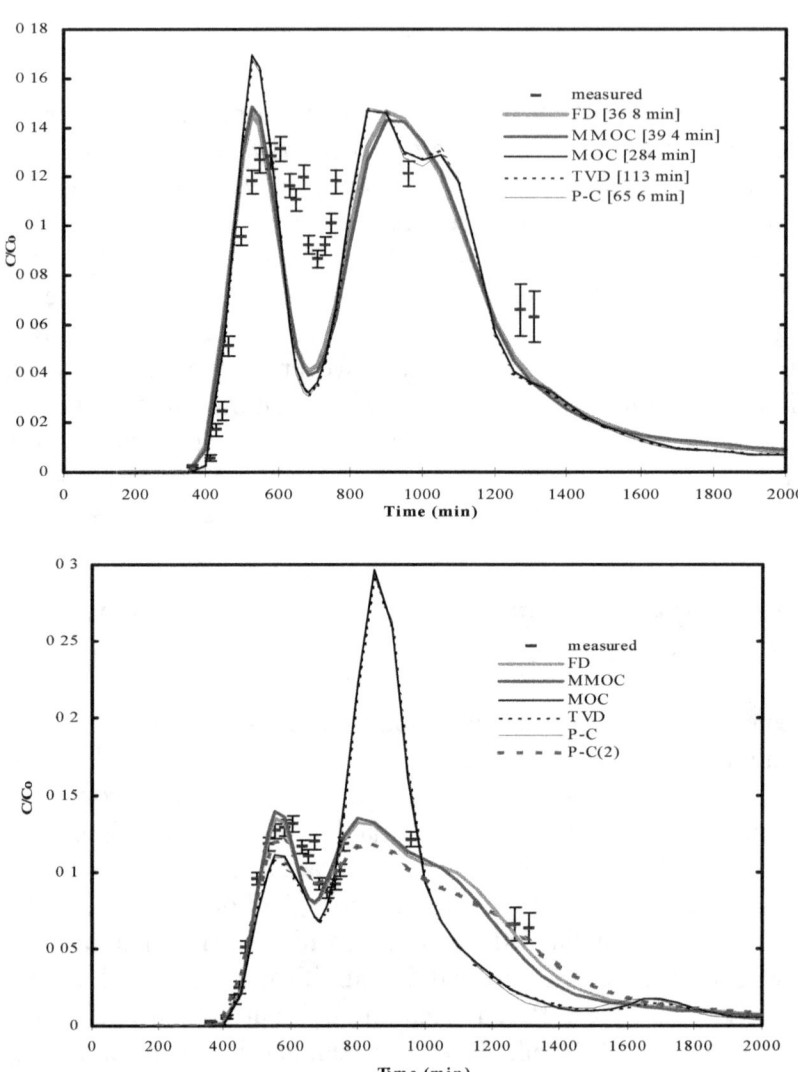

Figure 1. Results from Mehl and Hill (2001). For the measured concentration values, 95% confidence intervals are shown and reflect expected measurement error. Simulations used the finite-difference (FD), modified method of characteristics (MMOC), method of characteristics (MOC), and Total Variation Diminishing (TVD) numerical methods as coded in MT3DMS (Zheng and Wang, 1998), and a predictor-corrector (P-C) method coded for Mehl and Hill (2001). MOC, TVD, and P-C have the least numerical dispersion. (A) BTC's using measured hydraulic conductivities and dispersivities match measured concentrations poorly. Computation times are listed in brackets and are from a Linux workstation, Pentium II - 333, 64Mb Ram. (B) BTC's using optimized hydraulic conductivities and measured dispersivities. The solution labeled P-C(2) uses dispersivity values increased to approximate the numerical dispersion common to the FD and MMOC methods of MT3DMS.

4 USING A MODEL TO EVALUATE THE IMPORTANCE OF OBSERVATIONS TO PARAMETERS, PARAMETERS TO PREDICTIONS, AND OBSERVATIONS TO PREDICTIONS

Once a reasonably accurate simulation of a system has been achieved through careful model development, calibration, and error evaluation, the simulation itself becomes a valuable tool for sensitivity analysis, data assessment, and uncertainty evaluation. Sensitivity and data assessment methods can be categorized as identifying (1) observations that dominate model calibration (observations important to parameter values); (2) parameter values that dominate the predictions; and (3) observations that dominate the predictions. For instance, gradient-based methods such as dimensionless and composite scaled sensitivities, and parameter correlation coefficients (dss, css, and pcc); prediction scaled sensitivities, the value of improved information, and parameter correlation coefficients (pss, voii, and pcc); and the observation-prediction statistic (opr) can be used to address the three categories, respectively (Hill, 1998; Hill and others, 2001; Tiedeman and others, 2003). These local-sensitivity methods are often useful for nonlinear models, but can become useless if the nonlinearity is too extreme (Poeter and Hill, 1997; Hill, 1998). More computationally intensive methods that do not depend on model linearity include variance-based global sensitivity analysis methods for identifying parameters important to predictions, which address category 2 above (Saltelli and others, 2000), and jackknife and bootstrap methods for identifying observations that dominate parameter estimates or predictions, which address categories 1 and 3 (Davison and Hinckley, 1997)).

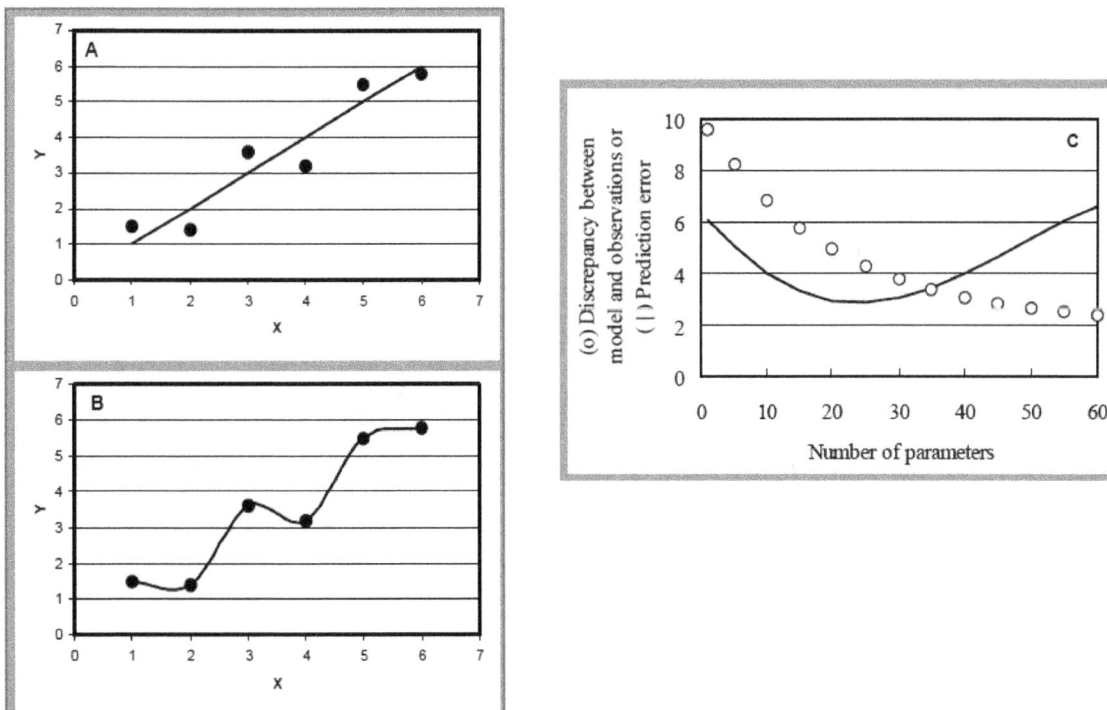

Figure 2. (A) Data with a true linear model. (B) The same data with an overly complex model with little predictive capability. (C) Schematic diagram showing the tradeoff between model fit to observations and prediction accuracy with an increasing number of parameters.

Figure **3** shows css values investigated by Barth and Hill (in review-b). The simulation mimics conditions of field experiments conducted by Schijven et al. [1999], and includes observations of hydraulic head (which have no sensitivity because the system is homogeneous and constant-head boundaries are imposed), flow through the system, normalized first temporal moments of conservative-transport concentrations, and virus concentrations. The observations provide the most information for the two hydraulic parameters, K and θ. Including TSS in Figure **3** allows evaluation of whether the information provided by the observations is sufficient to overcome typical numerical inaccuracies [*Barth and* Hill, in review-a]. Even with virus concentration observations, the css for λ1 is smaller than for TSS, suggesting that estimation of λ1 is likely to be affected by numerical inaccuracies.

5 CONCLUSIONS

Model development and evaluation are complex endeavors and predictions are always uncertain. To make wise societal decisions based on model predictions, it is important (1) for numerical methods to be as accurate as possible, and for any weaknesses to be sufficiently understood and accounted for; (2) to judge model fit in the context of a thorough evaluation of observation error; and (3) to use solid methods for evaluating the importance of observations to parameters, parameters to predictions, and observations to predictions.

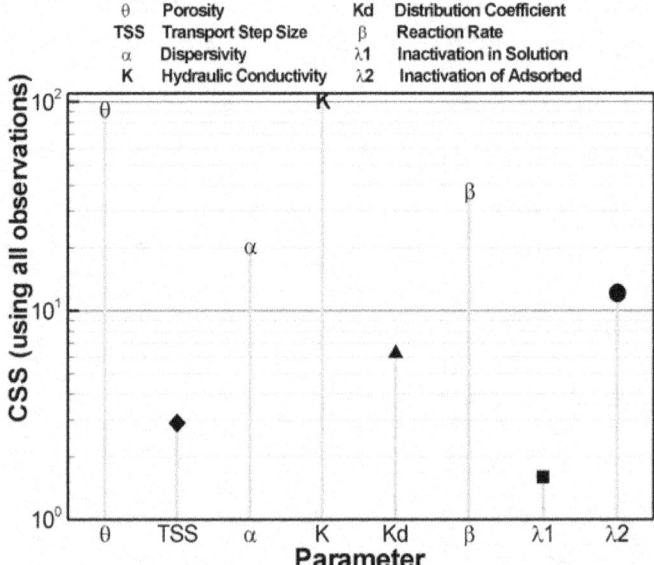

Figure 3. Composite-scaled sensitivities of seven system parameters and the simulation transport step size, TSS, evaluated using parameter set A. Observations include hydraulic heads, moments of conservative transport, and reactive transport concentrations. Composite-scaled sensitivities indicate the amount of information that the observations provide. K and θ are the most important parameters; TSS is more important than λ1. (from Barth and Hill, in review-b)

6 REFERENCES

Barth, G.R. and M.C. Hill, in review-a, Numerical methods for improving sensitivity analysis and parameter estimation of virus transport simulated using sorptive-reactive processes. journal article.

Barth, G.R. and M.C. Hill, in review-b, Parameter and observation importance in modeling virus transport in saturated systems – Investigations in a homogenous system: journal article.

Davison, A.C. and Hinckley, D.V., 1997, Bootstrap methods and their application. New York: Cambridge University press, 582p.

Hill, M.C., 1998, Methods and guidelines for effective model calibration. U.S Geological Survey Water-Resources Investigations Report 98-4005, 90p. Accessed 2/21/2004 at http://pubs.water.usgs.gov/wri984005/ .

Hill, M.C., Ely, M.D., Tiedeman, C.R., D'Agnese, F.A. Faunt, C.C., and O'Brien, B.A., 2001, Preliminary evaluation of the importance of existing hydraulic-head observation locations to advective-transport predictions, Death Valley regional flow system, California and Nevada. U.S. Geological Survey Water-Resources Investigations Report 00-4282, 82p, Accessed 2/21/2004 at http://water.usgs.gov/pubs/wri/wri004282/ .

Mehl, S.W. and Hill, M.C., 2001, A comparison of solute-transport solution techniques and their effect on sensitivity analysis and inverse modeling results. *Ground Water* 39(2): 300-307.

Poeter, E.P. and Hill, M.C., 1997, Inverse modeling, A necessary next step in ground-water modeling: Ground Water, 35(2), p. 250-260.

Saltelli, Andrea, Chan, Karen, and Scott, E. M., 2000, *Sensitivity Analysis*. John Wiley & Sons, NY, 475 p.

Schijven, J. F., W. Hoogenboezem, S. M. Hassanizadeh, and J. H. Peters, 1999. Modeling removal of bacteriophages MS2 and PRD1 by dune recharge at Castricum, Netherlands: *Water Resources Research* 35(4):1101-1111.

Tiedeman, C.R., Hill, M.C., D'Agnese, F.A., and Faunt, C.C., 2003, Methods for using groundwater model predictions to guide hydrogeologic data collection, with application to the Death Valley regional ground-water flow system: *Water Resources Research* 39(1): 5-1 to 5-17, 10.1029/2001WR001255.

Zheng, C. and P. Wang, 1998, MT3DMS- A modular three-dimensional multispecies transport model for simulation of advection, dispersion and chemical reactions of contaminants in groundwater systems, University of Alabama, Tuscaloosa.

4.3 Use of reactive-transport models in field studies: Experience with the PHAST simulator

D.L. Parkhurst and K.L. Kipp
U.S. Geological Survey, Denver, CO, dlpark@usgs.gov

1 INTRODUCTION

PHAST is a computer program for simulation of reactive-transport in three dimensions, which has been developed by the U.S. Geological Survey. The simulator is used to model constant-density ground-water flow, in confined or unconfined conditions, coupled with multi-component solute transport and a wide range of chemical reactions, including mineral, gas, ion-exchange, surface-complexation, and solid-solution equilibrium reactions and user-defined kinetic reactions. PHAST has been applied to field-scale studies of arsenic in the Central Oklahoma aquifer, aquifer storage and recovery (ASR) in Charleston, South Carolina, and phosphorus transport in a sewage plume at the Massachusetts Military Reservation (MMR), Cape Cod, Massachusetts. Development of the reactive-transport models for these studies has relied on extensive field and laboratory investigations. The PHAST simulations have aided in evaluating the pertinent transport and geochemical processes in the study areas and have provided predictive results that are being used in making water-resource-management decisions.

2 CAPABILITIES OF THE PHAST SIMULATOR

PHAST was developed by coupling the geochemical model PHREEQC (Parkhurst and Appelo, 1999) with a solute transport model derived from HST3D (Kipp, 1987; 1997). The simulator uses operator splitting to solve sequentially the coupled set of flow, transport, and chemical reaction equations, without iteration. This solution method is denoted as the sequential, non-iterative approach (SNIA) (Yeh and Tripathi, 1989). The advantages of SNIA are the simplicity of implementation and the ease of incorporation of complex chemical reactions. The numerical method has truncation errors arising from finite-difference discretization and operator splitting; but it appears possible to control these errors with sufficiently small spatial and temporal discretization.

PHAST uses an ion-association model to describe the thermodynamics of aqueous solutions. The reactions that can be modeled include homogeneous equilibrium within the aqueous phase and heterogeneous equilibria among the aqueous phase and minerals, gases, ion exchangers, surface complexation, and solid solutions. PHAST has a general method for kinetic reactions, where virtually any rate expression can be defined with Basic-language statements. The kinetic reactions are integrated either by a Runge-Kutta method or an implicit method. Both methods have automatic time-step adjustments to control errors.

The chemical calculations are the most computationally intensive part of reactive-transport simulations with PHAST. However, operator splitting between the chemical reaction step and the flow and transport steps leads to a natural parallelization of the calculations. The chemical reactions in each model cell are independent of all other cells and it is possible to distribute the workload to multiple processors with each processor calculating the chemistry for a subset of cells. The parallel version of PHAST distributes the workload of the chemical calculations by using the Message Passing Interface (MPI) (Parkhurst and Kipp, 2002). Randomization and reassignment of cells are used to balance the load among the processors. Using 30 processors, calculation time can be decreased by as much as a factor of 25.

The parallel version of PHAST makes it possible to run problems that have 100,000 nodes and 100s of time steps in a matter of hours on a cluster of Pentium IV computers. The parallel version was used to run field-scale simulations of Central Oklahoma aquifer, the Charleston ASR experiment, and the Cape Cod sewage plume.

3 NATURALLY OCCURRING ARSENIC IN THE CENTRAL OKLAHOMA AQUIFER

A 3-dimensional reactive transport model was developed to simulate the evolution of water chemistry in the Central Oklahoma aquifer over geologic time (10s of thousands of years) with special emphasis on large naturally occurring arsenic concentrations in the confined part of the aquifer, which limit the suitability of the aquifer for water supply. Model development relied on field investigations, including measurement of water levels and chemical analyses of ground water samples; and laboratory studies, including mineral identification by scanning electron microscope and X-ray diffraction, measurement of exchangeable cations in clays, and sequential extractions to identify the sorbed fraction of arsenic in the sediments (Parkhurst and others, 1996; Mosier and others, 1991).

The reactions for the reactive transport model include carbonate mineral equilibrium, cation exchange, and complexation of arsenic on sediment surfaces. The flow system was modeled as an unconfined aquifer in the east and a confined aquifer in the west. After 100,000 years of simulation, model results were consistent with field observations, showing calcium magnesium bicarbonate water in the unconfined aquifer and sodium bicarbonate water with large concentrations of arsenic in the confined aquifer. The simulations indicate that large concentrations of arsenic are likely to be a regional problem, which limits the utility of the aquifer as a water supply.

In detail, the simulations were not definitive. The largest concentrations of arsenic in the simulation did not extend as far south as is observed in the aquifer. This discrepancy between the model results and the observations can be attributed primarily to uncertainties in the representation of the flow system in the confined part of the aquifer.

4 AQUIFER STORAGE AND RECOVERY AT CHARLESTON, SOUTH CAROLINA

As part of an investigation of the possibility of storing potable water in a brackish aquifer for use during emergencies, a 3-dimensional reactive transport model was developed to consider the effects of chemical reactions in the aquifer on water chemistry and to estimate the amount of potable water that could be recovered from storage. The field investigation included two aquifer tests to determine hydraulic conductivity and aquifer storage, and four ASR experiments, where water chemistry at observation and recovery wells was measured at intervals during each experiment. X-ray diffraction was used to identify minerals and to estimate cation-exchange capacity of the aquifer sediments.

PHAST simulations indicated calcite equilibrium and cation-exchange reactions were sufficient to explain water-chemistry changes during the course of the ASR experiments. The most sensitive parameters for accurate simulation of the recoverable water were aquifer dispersivity coefficients, which described the dispersive transport in the system. After calibrating the dispersivities to match in the simulated and observed freshwater breakthrough at the observation wells, it was found that approximately one quarter of injected water could be recovered as potable water, regardless of the amount of water injected. This efficiency of recovery can now be used in the design of a full-scale ASR system.

5 PHOSPHORUS TRANPORT IN A SEWAGE PLUME AT THE MMR, CAPE COD, MASSACHUSETTS

Past disposal of sewage effluent through infiltration beds at a now-dismantled sewage treatment plant at the Massachusetts Military Reservation has generated a phosphorus plume that may potentially impact Ashumet Pond. The site has been extensively studied as part of the USGS Toxics Program and detailed information is known about porous media properties, ground-water velocities, and dispersivity in the aquifer (see for example, Garabedian and others, 1991; and LeBlanc and others, 1991). For the investigation of phosphorus transport, the key information for model development was delineation of the plume by analysis of water from many multilevel samplers, evaluation of sorption processes by laboratory column experiments with aquifer sediments, and investigation of oxygen consumption by degradation of organic matter accumulated in the sediments by laboratory batch and column experiments (Parkhurst and others, 2003).

Simulation of the flow system relied almost entirely on previous work at the site (including Garabedian and others, 1991; and LeBlanc and others, 1991). The chemical evolution of the plume was modeled with equilibrium reactions—surface complexation and reaction of iron and manganese oxide minerals—and kinetic reactions—organic decomposition and formation of iron phosphate minerals. Surface-complexation constants were derived by parameter estimation using a reactive-transport model of column experiments. The reactive-transport simulations indicated that loading of phosphorus to Ashumet Pond is near its peak currently and will decrease over the next few decades. The load estimates can be used in the process of developing a management strategy for the beneficial uses of Ashumet Pond. In this study, identification and parameterization of the chemical reactions was the greatest source of uncertainty.

6 CONCLUSIONS

The PHAST reactive-transport simulator has been developed by using SNIA for solution of the coupled flow, transport, and chemical reactions. A parallel version of the simulator has allowed simulation of field-scale problems in Oklahoma, South Carolina, and Massachusetts. The development of reactive transport models for these sites has relied on extensive field and laboratory investigations. Each study highlights data requirements for one of the three processes simulated in reactive transport modeling: the flow system in Oklahoma, dispersive transport in South Carolina, and chemical reactions in Massachusetts. The studies demonstrate that reactive-transport modeling is useful in understanding natural processes, in designing engineering projects, and in making management decisions for contamination issues.

REFERENCES

Garabedian, S.P., LeBlanc, D.R., Gelhar, L.W., and Celia, M.A. 1991. Large-scale natural-gradient tracer test in sand and gravel, Cape Cod, Massachusetts—2. Analysis of spatial moments for a nonreactive tracer. *Water Resources Research*: 27(5): 911-924.

Kipp, K.L. 1987. HST3D—A computer code for simulation of heat and solute transport in three-dimensional ground-water flow systems. *U.S. Geological Survey Water-Resources Investigations Report 86-4095*.

Kipp, K.L. 1997. Guide to the revised heat and solute transport simulator HST3D—Version 2. *U.S. Geological Survey Water-Resources Investigations Report 97-4157*.

LeBlanc, D.R., Garabedian, S.P., Hess, K.M., Gelhar, L.W., Quadri, R.D., Stollenwerk, K.G., and Wood, W.W. 1991. Large-scale natural-gradient tracer test in sand and gravel, Cape Cod, Massachusetts—Experimental design and observed tracer movement. *Water Resources Research*: 27(5): 895-910.

Mosier, E.L., Papp, C.S.E., Motooka, J.M., Kennedy, K.R., and Riddle, G.O. 1991. Sequential extraction analyses of drill core samples, Central Oklahoma aquifer. *U.S. Geological Survey Open-File Report 91-347*.

Parkhurst, D.L., and Appelo, C.A.J. 1999. User's guide to PHREEQC (Version 2)—A computer program for speciation, batch-reaction, one-dimensional transport, and inverse geochemical calculations. *U.S. Geological Survey Water-Resources Investigations Report 99-4259.*

Parkhurst, D.L., Christenson, Scott, and Breit, G.N. 1996. Ground-water-quality assessment of the Central Oklahoma Aquifer—Geochemical and geohydrologic investigations. *U. S. Geological Survey Water-Supply Paper 2357-C.*

Parkhurst, D.L., and Kipp, K.L., 2002, Parallel processing for PHAST—A three-dimensional reactive-transport simulator, In Hassanizadeh, S.M., Schlotting, R.J., Gray, W.H., and Pinder, G.F. (eds.), *Computational Methods in Water Resources, v. 2, Developments in Water Science.* 47, 711-718, Amsterdam, Elsevier.

Parkhurst, D.L., Stollenwerk, K.G., and Colman, J.A. 2003. Reactive-transport simulation of phosphorus in the sewage plume at the Massachusetts Military Reservation, Cape Cod, Massachusetts. *U.S. Geological Survey Water-Resources Investigations Report 03-4017*

Yeh, G.T., and Tripathi, V.S. 1989. A critical evaluation of recent developments in hydrogeochemical transport models of reactive multichemical components. *Water Resources Research.* 25: 93-108.

4.4 U.S. Environmental Protection Agency research and technical support needs related to conceptual model development for subsurface reactive transport modeling of inorganic contaminants, radionuclides, and nutrients

R.W. Puls[1]

[1]USEPA National Risk Management Research Laboratory, Ada, Oklahoma, puls.Robert@epa.gov

1 INTRODUCTION

The U.S. Environmental Protection Agency (USEPA) currently supports research through its Office of Research and Development (ORD) in the following high priority areas:

- Contaminated Sediments
- Ground Water Remediation
- Waste Characterization, Leaching, Reuse
- Arsenic
- Ecosystem Restoration
- Bioavailability (in support of risk assessments)

All of these areas utilize or could benefit from advancements in reactive transport modeling.

1.1 Specific areas of need for reactive transport modeling advancements

The following specific areas within the National Risk Management Research Laboratory (NRMRL) of USEPA/ORD can benefit from advancements in conceptual model development for subsurface reactive transport modeling of inorganic contaminants, radionuclides, and nutrients:

- performance assessment and site characterization for monitored natural attenuation of contaminants in the subsurface,
- performance assessment of reactive caps or monitored natural recovery for the treatment of contaminated sediments (in situ)
- conceptual models for prediction of contaminated sediment transport and contaminant bioavailability during resuspension events
- performance assessment of permeable reactive barriers for ground water remediation,
- improvements in speciation modeling, particularly for complex mixed waste sites, and
- improvements in conceptual model development for and modeling the transport of arsenic in subsurface systems and from wastes and waste products
- improvements in modeling the transport and fate of nitrogen in subsurface systems.

2   EPA/ORD CLIENTS

EPA's Office of Research and Development is responsible to EPA's Office of Solid Waste to provide research and technical support for waste site closures and the development of technical guidance in support of environmental regulations and programmatic policies. ORD is also responsible to EPA's regional offices to provide technical assistance for Superfund and RCRA cleanups. Many of these sites involve inorganic contaminants. Adequate site conceptual models are often lacking at many of these sites due to inadequate data, improper interpretation of data, lack of understanding and/or use of reactive transport models for site assessment purposes, and other factors.

2.1 Major sources of inorganic contaminants, radionuclides, nutrients as stressors for environmental programs

In terms of overall costs for remediation and restoration of sites impacted by inorganics (including radionuclides and nutrients), the most significant stressors (excluding air deposition) are Department of Energy (DOE) waste sites, abandoned mine lands, and agriculture. A recent top-to-bottom review of the DOE Environmental Management Program identified the following wastes and materials at DOE sites scattered around the country:

- 25 tons of plutonium
- 108 tons of plutonium residue
- 88 million gallons of radioactive liquid wastes
- 2500 tons of spent nuclear fuel
- 137,000 cu. meters of transuranic wastes
- 1.37 million cu. meters of low level wastes

The estimated life-cycle liability costs for managing these wastes exceed 200 billion dollars.

A recent EPA Office of Inspector General Report identified 156 hardrock mining sites nationwide that have the potential to cost between $7 billion and $24 billion to clean up (at a maximum total cost to EPA of approximately $15 billion). These costs are over 12 times EPA's total annual Superfund budget of about $1.2 billion for the last five years.

Another area of increasing focus is research on ecosystem restoration. While the majority of the focus here is on surface water quality, issues related to ground water – surface water interactions and sediment loading are also important. Non point source pollution from urban runoff, concentrated animal feeding operations, and agricultural practices have been identified as key stressors, with nitrogen management an increasingly important component of many restoration efforts. Nutrient management plans are often proving to be inadequate in controlling the transport of nitrogen, phosphate and other nutrients into receiving streams. It is possible that the conceptual models and transport models being used to assess nitrogen transport for these types of sites are inadequate.

The challenge confronting the USEPA, like many government agencies, is how to leverage scarce and shrinking resources to improve our capabilities in site conceptual model development and the development of improved reactive transport models for site assessment, remedial performance and watershed restoration efforts.

4.5 Reactive Transport Modeling of the Yucca Mountain Site, Nevada

G. S. Bodvarsson
Earth Sciences Division, Lawrence Berkeley National Laboratory

The Yucca Mountain site has a dry climate and deep water table, with the repository located in the middle of an unsaturated zone approximately 600 m thick. Radionuclide transport processes from the repository to the water table are sensitive to the unsaturated zone flow field, as well as to sorption, matrix diffusion, radioactive decay, and colloid transport mechanisms. The unsaturated zone flow and transport models are calibrated against both physical and chemical data, including pneumatic pressure, liquid saturation, water potential, temperature, chloride, and calcite. The transport model predictions are further compared with testing specific to unsaturated zone transport: at Alcove 1 in the Exploratory Studies Facility (ESF), at Alcove 8 and Niche 3 of the ESF, and at the Busted Butte site. The models are applied to predict the breakthroughs at the water table for nonsorbing and sorbing radionuclides, with faults shown as the important paths for radionuclide transport. Daughter products of some important radionuclides, such as ^{239}Pu and ^{241}Am, have faster transport than the parents and must be considered in the unsaturated zone transport model. Colloid transport is significantly affected by colloid size, but only negligibly affected by kinetic declogging (reverse filtering) mechanisms. Unsaturated zone model uncertainties are discussed, including the sensitivity of breakthrough to the active fracture model parameter, as an example of uncertainties related to detailed flow characteristics and fracture-matrix interaction. It is expected that additional benefits from the unsaturated zone barrier for transport can be achieved by full implementation of the shadow zone concept immediately below the radionuclide release points in the waste emplacement drifts.

4.6 Conceptual Model Development, Remediation, and Reactive Transport Modeling at Hanford: The U.S. DOE's Largest Legacy Waste Site

John M. Zachara[1], Chongxuan Liu[1], Jim McKinley[1], Zheming Wang[1], and Peter Lichtner[2]

[1]*Pacific Northwest National Laboratory, Richland, WA;* [2]*Los Alamos National Laboratory, Los Alamos, NM*

The Hanford site, which lies along the Columbia River in south-central Washington state, is the U.S. Department of Energy's largest legacy waste management site. The waste materials were derived from the reprocessing and extraction of Pu from irradiated U-fuels from 1940-1980. Hanford is located in a semiarid area, and has a variable thickness vadose zone (5-75 m) that overlies an unconfined aquifer that discharges to the Columbia River. Copious amounts of radioactive and chemical contaminants of highly variable mobility (e.g., ^{3}H, ^{129}I, ^{99}Tc, ^{60}Co, ^{90}Sr, ^{137}Cs, ^{234,235,236,238}U, ^{239,240}Pu, Cr) have been released to Hanford's vadose zone through intentional discharges of low concentration wastewaters to countless cribs, retention basins, and trenches; and accidental releases of concentrated high level wastes from massive underground storage tanks. Numerous vadose zone and groundwater plumes of different type and character exist. Some of these discharge directly to the Columbia River or threaten to do so in the future.

The site is currently undergoing one of the largest environmental remediation campaigns ever attempted. The scientific, engineering, and logistical challenges are enormous and complex. Remedial decisions are being based, in part, on projections of future in-ground behavior (e.g., reactive transport calculations) using models of different types and sophistication. In this presentation we will show examples of some of Hanford's more challenging and scientifically vexing environmental problems, and discuss conceptual geochemical and hydrologic models being developed to drive improved reactive transport calculations. The varied uses of reactive transport models will be identified as well as their data and information needs. Unavoidable uncertainties will be discussed that limit their more widespread use and acceptance. Recommendations will be provided on how to best integrate scientifically credible reactive transport modeling into remedial decisions at the complicated Hanford site.

4.7 NRC Modeling Applications and Needs

W.R. Ott, L.A. Veblen and J.D. Randall
US Nuclear Regulatory Commission, Washington , DC, WRO1@nrc.gov

1 INTRODUCTION

The US Nuclear Regulatory Commission (NRC) regulates all civilian uses of nuclear power and nuclear materials in the United States. Its primary mission is to protect the public health and safety and the environment from adverse effects from licensed activities. In carrying out this mission the NRC is faced with assessing the impacts of its actions. To assess these impacts it employs mathematical models, implemented in computer codes, of complex engineered and natural systems. Natural systems are modeled to estimate the migration of radioactive materials that have reached the natural environment through routine operational releases or because of accidental spills or other types of contamination (e.g. the mobilization of materials at an in situ leach uranium mine). Contaminated sites range from small, minimally contaminated areas that are easily remediated to large industrial processing sites with extensive surface and volumetric contamination and contaminated groundwater systems. The licenses for these sites cannot be terminated and the land released for unrestricted use until the potential doses to future users can be projected to meet the standards in the NRC's License Termination Rule (10 CFR 20, Subpart E). For disposal sites such as low-level waste disposal facilities and the proposed high-level waste (HLW) repository site at Yucca Mountain, NV, specific regulations govern their long term performance. The important concepts for this paper are that periods of regulatory interest extend from hundreds to thousands of years and concentrations of concern may be very small. Because we are dealing with radioactive species, phenomena such as radioactive decay and daughter in-growth in decay chains must be considered.

Performance assessment methodologies used in most early licensing and pre-licensing actions employed deterministic hydrologic models and constant distribution coefficients to account for sorption processes. We have known from the beginning that the simplifying assumptions inherent in these models were incorrect for most natural systems but limitations on computational systems made it impractical to do better. Although we accepted this limitation in a practical sense, we also acknowledged the need for modeling capabilities that could deal with more complex natural systems from both geochemical and hydrogeological perspectives. In particular, in order to be assured that public health and the environment received adequate protection, the NRC staff has consistently applied "conservative assumptions" and "bounding analyses" in a deliberate attempt to overestimate potential effects before applying regulatory criteria for guiding its decisions. With the substantial improvements in analytical capability that have evolved in the past thirty years the NRC has shifted its emphasis to more realistic assessments of the evolution of contaminated sites over time because it views unnecessary conservatism as a burden on licensees. The NRC's Office of Nuclear Regulatory Research is actively engaged in developing more realistic models to assess contaminated natural and anthropogenic systems.

2 THE SCOPE OF CURRENT EFFORTS

The current research program on radionuclide transport (NRC, 2002) at the NRC is approaching the problem of establishing more realistic waste-system performance models on several fronts. Natural heterogeneity in hydrologic systems has been addressed through a systematic effort to develop probabilistic techniques to accommodate parameter uncertainty, conceptual model uncertainty, and scenario uncertainty (Meyer and Orr, 2002; Neuman and Wierenga, 2003). This work has been focused at the University of Arizona and the Pacific Northwest National Laboratory (PNNL). The limitations of off-the-shelf models have led to support of developmental work on models implemented in software that can combine and interface computational modules to represent discrete subsystems. The current focus for this work, in cooperation with the Department of Energy, the Environmental Protection Agency, and the U.S. Army Corps of Engineers is the FRAMES modeling system being developed at PNNL. Simpler models have not been abandoned because most of sites that the NRC regulates have minimal contamination in a relatively confined area and are thus compatible with such modeling systems. However, even these are being improved with probabilistic capabilities and improved databases to support selection of parameter values and to move away from the use of default values. A third area of recent study has been model abstraction techniques to help focus model definition with respect to actual site data.

The last area to mention here is the area of reactive transport. Since the late 1980's the NRC has been supporting work on sorption in one form or another. Early efforts focused on the potential HLW repository at Hanford, Washington, and were focused on the distribution coefficient measurements at elevated temperatures for the rock systems and radionuclides of interest (Apps et al., 1989). This work evolved into the geochemistry program at the Center for Nuclear Waste Regulatory Analyses which has made significant advances in modeling sorption processes for the NRC performance assessment code for the current potential HLW repository at Yucca Mountain (Turner, 1993, Pabalan et al. 1998, Prikryl et al., 2001). Later talks will discuss this work in some detail. The non-HLW research at the NRC has focused on ambient temperature systems which will still be relevant in the far field of the repository but are also more relevant to decommissioning of nuclear facilities and near-surface disposal facilities. The early focus was on the natural analogue studies at the Koongarra Ore Body in northern Australia (Duerden et al., 1992) and involved the U.S. Geological Survey (USGS) as our principal contractor (Davis. 2001). After two multinational international projects studying uranium migration at this hydrologically and geochemically complex site, our principal investigator concluded that semi-empirical surface complexation models could be used to develop more realistic models of uranium migration at a chemically complex site. We then funded a demonstration project with the USGS at the Naturita uranium remediation site in Colorado in the United States. Most of the work at that site has been completed and a final report has been issued (Davis and Curtis, 2004). We are now working to broaden these techniques for use at a wider range of sites and a wider range of radionuclides. Our interest in this topic has led us to be participants in both the first and second phases of the multi-national NEA Sorption Project (NEA, 2000) which is nearing the completion of its second phase. This project has been testing the modeling approaches used to describe sorption by modelers in the participating countries. The last and perhaps most intriguing part of our

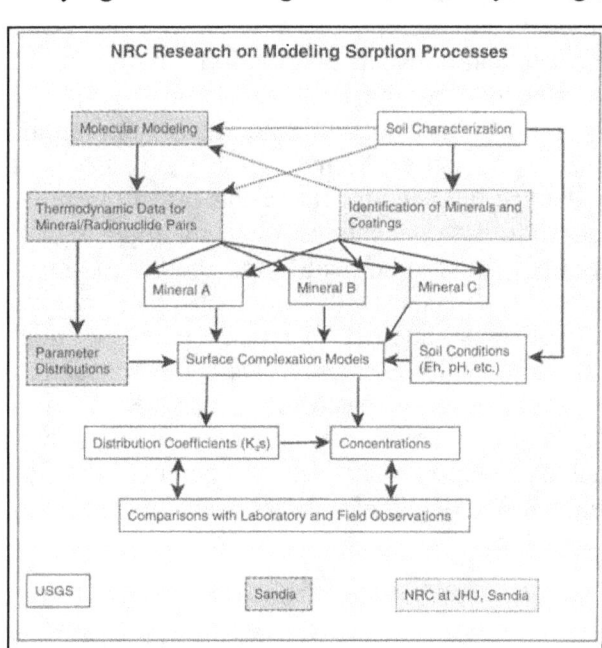

Figure 4: The NRC sorption research program, showing roles of the USGS, Sandia, and the NRC staff

program is that being carried out at Sandia National Laboratories. Part of this program has been supporting characterization of sorbing minerals at the Naturita site in suppport of the USGS team (McLain et al., 2002, and Jové-Colon et al., 2004). Other parts of this research include (1) assessing the effects of adsorption constant uncertainty on contaminant plume migration through one and two dimensional studies, and (2) modeling sorption processes at the molecular level (Cygan, 2002, Teter and Cygan, 2002, Cygan et al., 2004). Figure 1 presents a crude flow chart for how this work might be integrated with field measurements. It has been, and remains, our goal to connect this work and produce a technically credible, scientifically sound basis for the semi-empirical models.

3 SUMMARY AND RELATION TO WORKSHOP

The NRC is not the only federal agency that uses environmental models to inform its decisions. It has been evident for many years that expertise and parallel research programs existed in several federal agencies. Bilateral cooperation on individual projects happened but was usually not sustained. About three and a half years ago a group of federal researchers met to discuss the difficulties of exchanging data and computational modules between programs. That interagency workshop led to the idea that a more permanent structure would be a real benefit to all of the agencies involved. Two and half years ago the Memorandum of Understanding under which this workshop is being held became effective with the signature of the last of six original participating organizations. This is the third working group formed under the MOU to reach the stage of sponsoring an international workshop to discuss opportunities and new approaches for an important area of scientific investigation related to the improvement of multimedia environmental models. As one of the organizers and a member of the MOU's Steering Committee, we look forward to the further deliberations of this workshop and the fruits of continued cooperation among the participating agencies.

REFERENCES

Apps, J.A., McNeil, J.M., and Jun, C.-H. 1989. *Thermochemical Properties of Gibbsite, Bayerite, Boehmite, Diospore, and the Aluminate Ion between 0 and 350 degrees C*, NUREG/CR-5271, US Nuclear Regulatory Commission, Washington, DC.

Cygan, R. T., 2002, Molecular Models of Radionuclide Interaction with Soil Minerals, in Zhang, P.C, and Brady, P. V. (eds.), *Soil Science Society of America Special Publication Number 59, Geochemistry of Soil Radionuclides,* Madison, Wisconsin.

Cygan, R.T., Liang, J.-J., and Kalinachev, A.G. 2004. Molecular Models of Hydroxide, Oxyhydroxide, and Clay Phases and the Development of a General Force Field, *Journal of Physical Chemistry B,* in press (2004).

Davis, J.A. 2001. Surface Complexation Modeling of Uranium(VI) Adsorption on Natural Mineral Assemblages, NUREG/CR-6708, US Nuclear Regulatory Commission, Washington, DC.

Davis, J.A., and Curtis, G.P. 2004. Application of Surface Complexation Modeling to Describe Uranium(VI) Adsorption and Retardation at the Uranium Mill Tailings Site at Naturita, Colorado, NUREG/CR-6820, US Nuclear Regulatory Commission, Washington, DC.

Duerden, P., Lever, D.A., Sverjensky, D.A., and Townley, L.R. 1992. Alligator Rivers Analogue Project Final Report, Volume 1, Summary of Findings, ISBN 0-642-59927-0, Organization for Economic Cooperation and Development, Nuclear Energy Agency, Paris, France.

Jové-Colon, C.F. et al., 2004. A Combined Analytical Study to Characterize Uranium Soil Contamination: The Case of the Naturita UMTRA Site and the Role of Grain Coatings, US Nuclear Regulatory Commission, Washington, DC.

McLain, A. A., Altman, S. J., Rivers, M. L., and Cygan, R. T., 2002. Use of Computerized Microtomography to Examine the Relationships of Sorption Sites in Alluvial Soils to Iron and Pore Space Distributions, NUREG/CR-6784, US Nuclear Regulatory Commission, Washington, DC.

Meyer, P.D., and Orr, S. 2002. *Evaluation of Hydrologic Uncertainty Assessments for Decommissioning Sites Using Complex and Simplified Models*, NUREG/CR-6767, US Nuclear Regulatory Commission, Washington, DC.

NEA 2000. Using Thermodynamic Sorption Models for Guiding Radioelement Distribution Coefficient (K_d) Investigations for Performance Assessment of Repositories for Radioactive Wastes, OECD Publications, 2, rue André-Pascal, 75775 Paris Cedex 16, France (2000). Also, see http://www.nea.fr/html/jointproj/sorption.html.

Neuman, S.P., and Wierenga, P.J. 2003. A Comprehensive Strategy of Hydrogeologic Modeling and Uncertainty Analysis for Nuclear Facilities and Sites, NUREG/CR-6805, US Nuclear Regulatory Commission, Washington, DC.

NRC 2002. Radionuclide Transport in the Environment: Research Program Plan, March 2002, US Nuclear Regulatory Commission, Washington, DC, also available at http://www.nrc.gov/reading-rm/adams/web-based.html, ADAMS Accession Number ML020660731.

Pabalan, R.T., Turner, D.R., Bertetti, F.P., and Prikryl, J.D. 1998. Uranium(VI) Sorption onto Selected Mineral Surfaces: Key Geochemical Parameters. In E.A. Jenne (ed.), Adsorption of Metals by Geomedia: 99-130. San Diego, CA: Academic Press.

Prikryl, J.D., Jain, A., Turner, D.R., and Pabalan, R.T. 2001. Uranium(VI) Sorption Behavior on Silicate Mineral Mixtures, Journal of Contaminant Hydrology, 47, 241-253.

Teter, D.M., and Cygan, R.T. 2002. Large-Scale Molecular Dynamics Simulations of Metal Sorption onto the Basal Surfaces of Clay Minerals, NUREG/CR-6757, US Nuclear Regulatory Commission, Washington, DC.

Turner, D.R. 1993. Mechanistic Approaches to Radionuclide Sorption Modeling, CNWRA 93-019, Center for Nuclear Waste Regulatory Analyses, San Antonio, TX.

4.8 Predicting Soil and Water Chemistry in and below the Root Zone of Agricultural Lands: Major Ions, Nutrients and Toxic Ions

D.L. Suarez

USDA-ARS, Salinity Laboratory, Riverside, CA dsuarez@ussl.ars.usda.gov

The focus of ARS research on inorganic chemicals centers around the need to maintain a productive soil environment for crop production while minimizing or avoiding off-site degradation of soil and water resources. The inorganic chemicals of interest are salts, nutrients and potentially toxic ions. Modeling is necessary for predict of the leaching of chemicals under existing practices and more importantly as a management tool to evaluate new practices. Evaluation of practices must of course consider not only chemical discharge but crop productivity and sustainability.

The environment of primary interest is the soil root zone. In some instances groundwater and surface water processes and transport is also of interest. The root zone environment is a dynamic region of the soil with changes in water content, plant uptake of water and chemicals, changing redox and gas phase composition and in many instances short residence times. Chemical modeling is highly dependent on an accurate description of water flow, thus a variably saturated water flow routine is often needed. Plant water uptake is an important component of water flow and determinant of solute composition (evapotranspiration concentrates salts into the remaining soil water). In most instances chemical modeling in the root zone requires modeling of plant water uptake in response to environmental stresses (water, salts, ions, nutrients, and climate). Since the system is dynamic, predictions based on equilibrium (thermodynamic) assumptions will usually not be satisfactory. Additional nutrients (NH_4, PO_4 and B), and toxic elements (Se, As, B, and Mo, as well as some heavy metals).

1 MODELING MAJOR ION CHEMISTRY

1.1 *Rationale and processes*

Modeling of major dissolved ions (Na, K, Ca, Mg, Cl SO_4, NO_3, and HCO_3 +CO_3) is necessary for salinity control, primarily for western irrigated agriculture. High salt concentrations reduce crop yield and water consumption. In this instance a plant stress response model is needed to model salt composition and transport. Major ion processes needed to be modeled include mineral precipitation-dissolution, and cation exchange. Minerals of major importance are calcite and gypsum. Since calcite solubility depends on CO_2 carbon dioxide partial pressure, there is a need to simulate gas concentrations. In some instances it may be sufficient to estimate annual or seasonal gas phase profiles with a fixed boundary condition, in other instances CO_2 dynamics are important and both production and transport are necessary to predict concentration. Calcite kinetics may need to be considered in some instances or alternatively the mean ion activity product for calcium carbonate is relatively stable for short residence times and is a better predictor than calcite equilibrium (Suarez et al., 1992). Precipitation dissolution of more soluble salts such as mirabilite and halite and generally not necessary since in an agricultural setting these salts would only occur in a dry surface crust and quickly resolubilize after irrigation or rainfall.

1.2 Leaching and salinity control

Salinity control in the root zone is a major concern for western irrigated agriculture. Salt leaching must consider soil mineralogy, hydraulic properties, climate, and plant water uptake. Salinization may occur as a result of under irrigation and/or capillary rise from shallow saline ground water. Plant response to salts is snot just related to overall salt composition, but to specific ion concentrations as well, thus need be to predict individual ion concentrations.

1.3 Sodic soil reclamation

High levels of exchangeable Na, especially when salinity levels are low, results in adverse soil physical properties, including reduction in hydraulic conductivity, adversely affecting water infiltration, aeration and soil tilth. The UNSATCHEM code simulates the effects of pH, salinity and exchangeable Na percentage on hydraulic conductivity, utilizing a modification of the variable saturated water flow equation as follows (Suarez and Simunek, 1996)

$$K(h) = K_s \, K_r \, r = K_s \, r \, S_e^{1/2}[1-(1-S_e^{1/m})^m]^2 \tag{1}$$

where $K(h)$ is the hydraulic conductivity at the specified pressure head, K_s is the saturated hydraulic conductivity, K_r is the relative hydraulic conductivity (scaled), r is the reduction function defined by the chemistry of the soil solution, S_e is defined by $(\theta_w - \theta_r)/(\theta_s - \theta_r)$, and m is an empirical parameter. The θ_w θ_r and θ_s correspond to the existing, residual, saturation water contents respectively. This is an important consideration for prediction of water flow in saline soils as shown by the results of column experiments, Fig 1 (Suarez et al.1984)

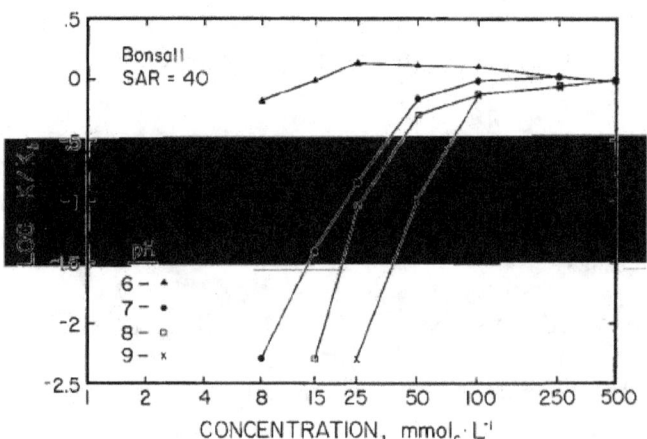

Figure 1 Change in relative saturated hydraulic conductivity as related to solution concentration, SAR and pH.

Figure 2a shows a comparison of predicted versus observed effect of gypsum mixed into the top 15 cm of a saline sodic soil subsequently leached with dilute water (Suarez, 2001). The SAR (sodium adsorption ratio) is directly related to the exchangeable Na percentage (percentage of the cation exchange sites that are occupied by Na). Under enhanced CO_2 production and reduced transport in the gas phase, calcite can be an appreciable source of calcium, suitable for reclamation. The predictions shown in Fig 2b match field observations of reclamation of calcareous soils using this "green manuring" technique.

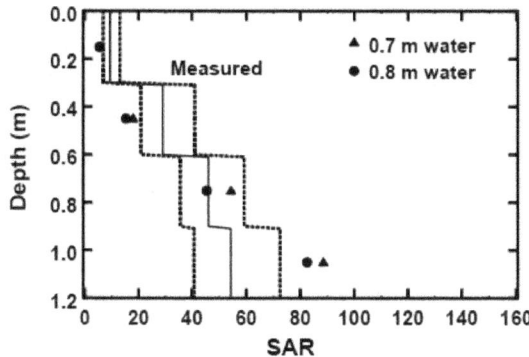 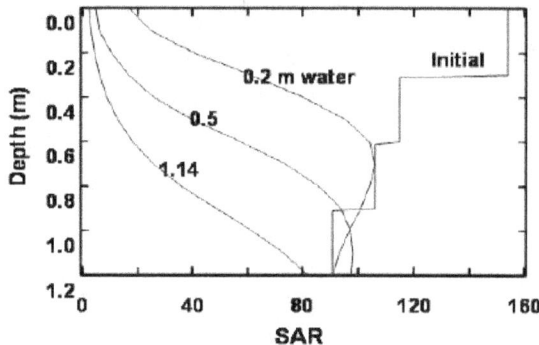

Figure 2 a) Comparison between predicted and observed SAR with depth after application of gypsum and leaching with 0.74 m of water(and b) Predicted change in SAR as related to water applied, utilizing calcite dissolution and management practices to enhance CO_2 concentration in the soil.

2 NUTRIENTS

2.1 *Nitrogen*

Nitrate leaching from agricultural soils is a major environment concern. High crop productivity requires high N availability. The RZWQM (Ahuja et al. 1999) contains a chemical routine for cation exchange including NH_4 and N transformations from nitrate, NH_4 and various organic matter N pools in the soil, as well as consideration of ammonia volatilization. ARS needs in this area are primarily to predict the concentration of nitrate leached below the root zone as related to nitrogen applications and management practices (tillage, fallowing fertilizer timing etc.), however knowledge of methane production and losses to the atmosphere are also needed to evaluate greenhouse gas production.

Many nutrient models utilized by ARS do not attempt to model the chemical reactions but rather provide regression type routines to predict nutrient losses. These models are useful but not within the scope of the present topic, process- based chemical modeling.

2.2 *Phosphorous*

In agricultural environments, transport of P is primarily a concern associated with sediment transport, thus mostly associated with physical erosion models rather than chemical reactions. Adsorption/precipitation of P is of interest in as much as it impacts plant P uptake.

3 TOXIC IONS

Saline water is often associated with elevated concentrations of B, Se, As and Mo. These elements adsorb to mineral surfaces thus prediction of their concentrations requires adsorption modeling. For Se and As there is also a need to consider redox reactions, volatilization of organic gases and incorporation into organic matter. Chemical modeling is necessary to predict transport. Boron is both an essential plant nutrient and toxic at higher concentrations. Modeling of B adsorption/desorption and transport has been simulated using UNSATCHEM (Suarez and Simunek, 1996.) Figure 3 shows the predictive capability of the model utilizing the constant capacitance model for adsorption and generalized model parameters (Suarez, 2002).

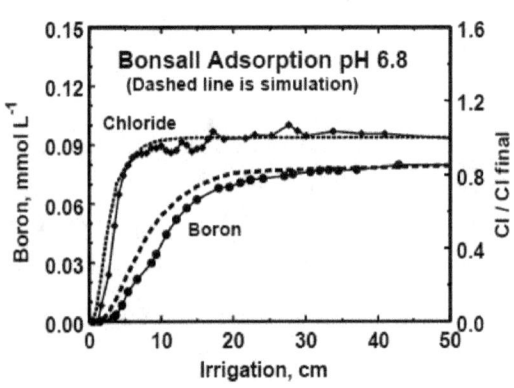 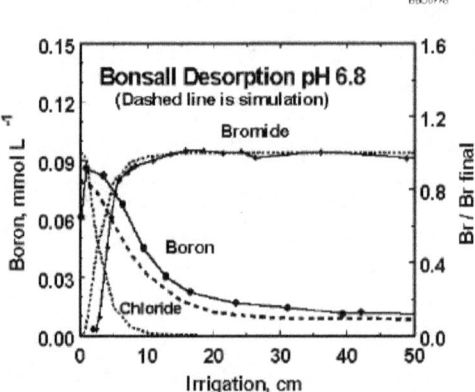

Figure 3 Comparison of UNSATCHEM predictions and column data for a) Cl and B leaching during adsorption and b) CL and B leaching during desorption.

4 HEAVY METALS

Heavy metals are of concern in urban environments as well as former orchards where lead arsenate was used as an insecticide. At present, transport modeling has not been utilized, but would be useful for evaluation of remediation options.

5 REFRENCES

Ahuja, L.R., Rojas, K.W. Hanson, J.D., Shaffer, M.J, & Ma, L. 1999. Root zone water quality model. Highland Ranch: Water Resources Publications, LLC.

Suarez, D.L. 2001. Sodic soil reclamation: Model and field study. *Aust. J. Soil Res.* 39:1225-1246.

Suarez, D.L. 2002. Evaluation of management practices for use of low quality waters for irrigation: Model simulations. In, *Proceedings of the international union of soil scientists. August 14-21, 2002.* 1096:1-8. Bangkok.

Suarez, D.L., Rhoades, J.D., Lavado, R., & Grieve, C.M. 1984. Effect of pH on saturated hydraulic conductivity and soil dispersion. *Soil Sci. Soc. Am. J.* 48:50-55

Suarez, D.L. & Simunek, J. 1996. UNSATCHEM code for simulation one-dimensional variably-saturated water flow, heat transport, carbon dioxide production and transport, and multi component solute transport with major ion equilibrium and kinetic chemistry. *U.S. Salinity Laboratory Tech. Report. No. 129.*

Suarez, D.L. & Simunek, J. 1997. UNSATCHEM: Unsaturated water and solute transport mode with equilibrium and kinetic chemistry. *Soil Sci. Soc. Am. J.* 61:1633-1646.

Suarez, D.L, Wood, J.D., & Ibrahim, I. 1992. Reevaluation of calcite supersaturation in soils. *Soil Sci. Soc. Am. J.* 56:1776-1784.

4.9 ERDC Activities and Needs Related to the Reactive Transport Modeling of Inorganic Contaminants

C.J. McGrath

[1]*U.S. Army Engineer R&D Center, Vicksburg, MS, Chris.McGrath@erdc.usace.army.mil.*

The U.S. Army Engineer Research and Development Center (ERDC) consists of seven laboratories, which collectively serve the U.S. Army Corps of Engineers, The Department of the Army, and other Federal, State, and international organizations. The ERDC conducts basic and applied environmental research and manages projects to address both military and civilian priorities. The vast majority of project and research activity involves organic compounds, *e.g.*, explosives and energetics in support of military needs, and PAHs, PCBs, solvents, and fuels in the civil works arena. However, heavy metals from both military activity (*e.g.*, munition-related Pb, DU) and civil projects (contaminated sediments and dredging) pose environmental concerns. Military environmental research is focused on military-unique issues such as installation restoration, training range maintenance, and minimizing impacts of military actions. A wide variety of common inorganic and organic contaminants are encountered on active, former, and realigning military installations.

ERDC supports the development and modification of models describing the reactive transport processes influencing the natural fate and engineered cleanup of contaminants. One of the centerpieces of the ERDC modeling program is a suite of graphical user interfaces (GUIs) – the Groundwater Modeling System (GMS), Surface-water Modeling System (SMS), and Watershed Modeling System (WMS) – each interfacing with several popular models, supplemented with system-specific tools that enhance model application. These GUIs are available commercially or at no cost to supporting Federal agencies and their contractors, *e.g.*, the GMS is supported by EPA, DOE, and NRC. Future model development supported by ERDC will tend to utilize xMS interfaces and models. The GMS will continue to evolve in directions supported by evolving project needs and research.

4.10 Direct Visualization of Reactive Transport: Mixing at the Pore-Scale and at the Darcy-Scale

P. Oates[1], C. F. Harvey[1], L. Meigs[2], and R. Haggerty[3]

[1]*Parsons Lab, CEE, MIT, Cambridge, MA, charvey@mit.edu*

[2] *Sandia National Labs, Albuquerque, NM*

[3] *Dept. Geoscience, Corvalis, OR*

We have designed laboratory experiments to study two basic problems in hydrogeology: (1) How do chemicals mix so that they react in natural porous media and, (2) What model best describe solute spreading in the subsurface? These questions are intimately related because Darcy-scale spreading, pore-scale mixing, and chemical reaction may profoundly affect one another.

We have developed methods that accurately image changing concentrations during reaction in porous media experiments using colorimetric chemical reactions. These methods enable us to distinguish product from reactants and quantify their concentrations throughout experimental porous media chambers by their different light absorption (e.g. a green product formed from clear reactants). The results show that the conventional coupling of chemical reaction equations with transport equations developed for conservative solutes may either over predict or under predict the degree of chemical reaction. In some cases, incomplete mixing, at both the pore-scale and the Darcy-scale (from cm size heterogeneity), segregate reactants so that the rate of reaction is over predicted by conventional coupling of transport and chemical reaction equations. In other cases chemical reaction creates density fingering, further mixing the reactants, and greatly enhancing reaction. Density gradients develop under conditions of slow flow because diffusion coefficients differ between reactants and products. We will discuss how processes that mix or segregate solutes depend on the spatial texture of hydraulic conductivity.

4.11 Modeling Fate and Transport of Inorganic-Organic Co-Contaminants at a DOE Fractured Waste Site

J.-P. Gwo[1] and P.M. Jardine[2]

[1]*University of Maryland, Baltimore County, Baltimore, MD, jgwo@umbc.edu.*
[2]*Oak Ridge National Laboratory, Oak Ridge, TN*

1 ABSTRACT

The objective of the research is to study the effects of physical and chemical heterogeneities at matrix block and field scales on the fate and transport of Co(II)EDTA at a DOE low-level waste disposal site at Oak Ridge National Laboratory (ORNL). The field site, Waste Area Grouping 5 (WAG 5), is highly heterogeneous with confined, fractured aquifers and partially weathered shale bedrocks and saprolite near ground surface. Laboratory experiments using undisturbed saprolite soil columns and field experiments using multiple reactive and non-reactive tracers have been conducted to elucidate mechanisms governing the fate and transport of metal and metal-EDTA chelates such as Cr, Co(II)EDTA, CdEDTA and non-reactive tracers such as Br, He, and Ne. Laboratory observations from Co(II)EDTA injection into a saprolite soil was used to calibrate a two-pore-domain, fracture-matrix reactive transport model. The goal of the study is to determine mass transfer and transport parameters and kinetic reaction rate coefficients that characterize physical and chemical non-equilibrium, respectively, at matrix block scale. Model parameters thus obtained were then used for a field scale transect model of WAG 5 to study the effect of field scale heterogeneity on the fate and transport of Co(II)EDTA. The field scale model was first calibrated hydraulically using field observations of groundwater levels at monitoring wells installed along the transect. Tracer breakthrough curves at these wells were then used to calibrate a two-pore-domain model for the transport of non-reactive tracers (Br and Ne). The latter modeling effort was designed to study the effects of physical non-equilibrium on field scale transport of non-reactive tracers. Finally, sensitivity analysis is conducted to study the effects of physical and chemical heterogeneities on the fate and transport of Co(II)EDTA at field scale.

2 INTRODUCTION

Matrix diffusion is a widely observed phenomenon in fractured porous medium (e.g., Neretnieks, 1980; Novakowski and Lapcevic, 1994; Jardine et al., 1999). It is a comparatively small-scale effect that has also been observed to have effects on field scale movement of solutes (e.g., Jardine et al., 1999). Previous modeling studies of fate and transport in fractured geological formations include equivalent porous medium (e.g., Lee et al., 1992; Allen and Michel, 1999; Bodvarsson et al., 2003), discrete single or multiple fractures of various shapes and connectivity (e.g., Sudicky and McLaren, 1992; Tsang and Tsang, 1987), and models consisting of two or more interactive pore domains (e.g., Gerke and van Genuchten, 1993; Gwo et al., 1996). These models and their computational counterparts, from simplistic to complex, have various degrees of utility as more and more information of the subsurface is gathered. However, large uncertainty may exist because it is often impractical to characterize experimentally field scale heterogeneities at matrix block scale (Gwo et al., 2004b). For reactive transport of metal and metal-chelates, this observation is further complicated by the lack of information in solid surface properties that may vary from pore domain to pore domain and from formation to formation. As a result, characterization and remediation of contaminated field sites may require more extensive, and often times more expensive, data collection and analysis efforts. Sensitivity and cost-benefit analyses may also be necessary to address the issues of characterizing physicochemical heterogeneity, prioritizing data collection efforts and managing monetary risks of various remediation designs (e.g., James et al., 1996).

The objective of the research is to study the effects of physical and chemical heterogeneities at matrix block and field scales on the fate and transport of Co(II)EDTA at fractured waste sites. A multiscale approach involving laboratory and field injections of Co(II)EDTA into undisturbed saprolite soil column and fractured formations at the WAG 5 site, respectively, is used. Reactive transport model parameters obtained from the calibration of a soil column model are used in sensitivity analysis for a field scale, transect model of WAG 5. The goal is multifold: (1) determine the most sensitivity parameters for field scale model calibration effort using field observations of reactive tracer injection (Jardine et al., 2002), (2) determine the relative effects of physical and chemical mass transfer, and (3) determine the most relevant scales of various physicochemical processes.

3 METHODOLOGY

A transect of one source well, two multilevel monitoring wells, and 20 clusters of drive-point wells were installed at the WAG 5 site (Figure 1). Non-reactive tracers, metal and metal-EDTA chelates were released in two successive experiments at the source well. Bromide, helium and neon were released in the first experiment to study the effect of matrix diffusion on field scale transport and to characterize physical non-equilibrium as a result of tracer movement between pore domains or fracture and matrix. Metal and metal-EDTA chelates were released in the second experiment to study their fate and transport in highly structured soils and geological formations. Groundwater levels and movements of the chemicals were monitored through the multilevel and drive-point wells. Undisturbed soil columns were collected from soil horizon C, mostly saprolite that is partially weathered shale bedrock and may have retained the majority of the physical and chemical properties of the bedrock. Co(II)EDTA tracer injection was conducted in the laboratory using the undisturbed soil columns.

A two-pore-domain, fracture-matrix groundwater flow model is calibrated using observed groundwater level and a multiple-pore-domain flow simulator MURF (Gwo et al., 1994; Gwo et al., 2004a). Flow fields in the fracture and matrix pore domains thus calculated are used to calibrate a non-reactive transport model of Br and He release at WAG 5 (Gwo et al., 2004b), using

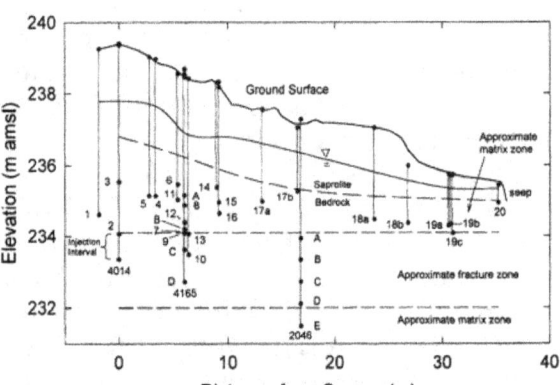

Figure 1. ORNL WAG 5 transect and wells.

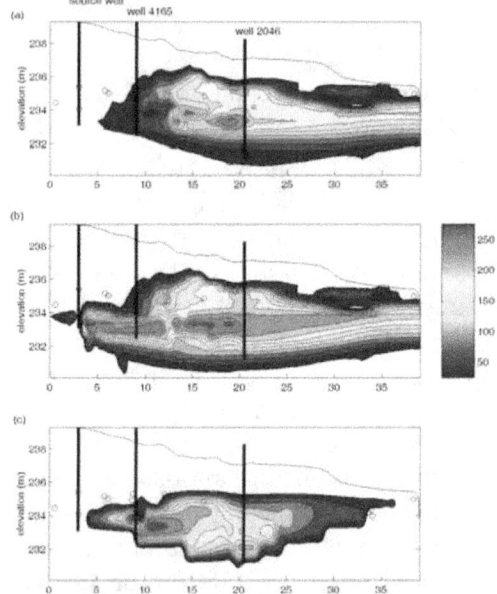

Figure 2. Observed Br concentration (c) and those calculated in fracture and matrix (a and b, respectively).

the tracer release data and a multiple-pore-domain transport simulator MURT (Gwo et al., 1995; Gwo et al., 2004b). The purpose of the calibration efforts is to determine advective and diffusive mass transfer coefficients that characterize physical non-equilibrium as results of pressure and concentration differences between pore domains, in addition to determining field scale heterogeneities that are responsible for preferential flow and transport of fluids and solutes. Shown in Figure 2 are the calculated Br plumes in the fracture and matrix (panels a and b) and the observed Br plume (panel c).

4 RESULTS AND DISCUSSION

A network of interfacial reactions were used to represent the surface MnO_2 mediated transformation of Co(II)EDTA to Co(III)EDTA. Laboratory observations are used to determine kinetic reaction rate coefficients for the oxidation of Co(II)EDTA to Co(III)EDTD by surface MnO_2, by fitting model predictions to observed soil column tracer breakthrough curves (Figure 3). The results indicate that both physical and chemical mass transfer processes may affect the adsorption and movement of CoEDTA in structured soils.

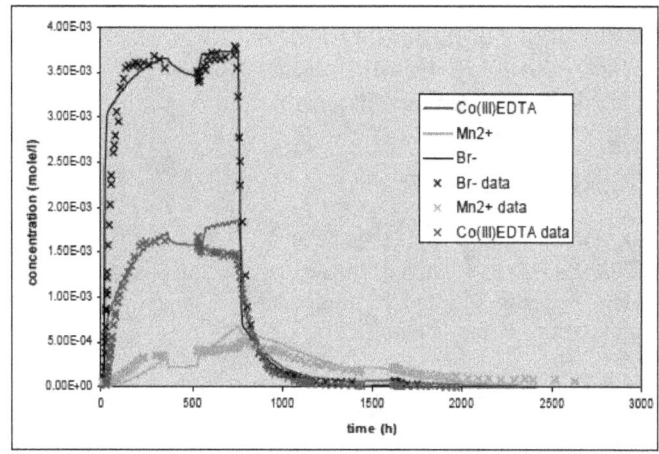

Figure 3. Predicted and observed reactive tracer breakthrough curves.

The reaction network, the kinetic rate coefficients obtained from the soil column modeling exercise and the physical mass transfer coefficients obtained from the field scale model of non-reactive tracer transport are subsequently used in modeling field scale fate and transport of Co(II)EDTA. Large scale heterogeneities, including matrix and fracture characteristic length, porosity, dispersivity, tortuosity and flow fields of pore domains obtained from previous flow and non-reactive solute transport model calibrations are

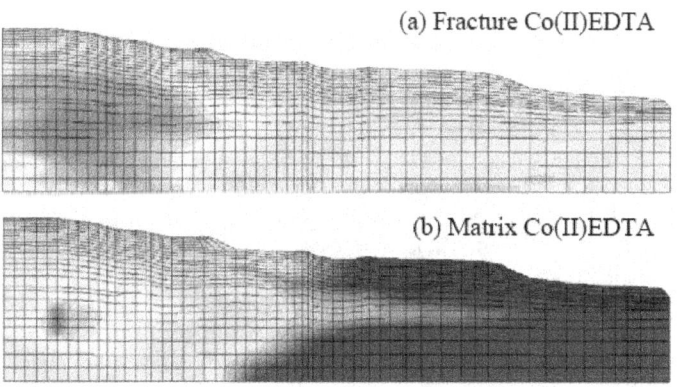

Figure 4. Predicted Co(II)EDTA concentrations in the (a) fracture and (b) matrix. Note that concentrations are plotted in logarithmic scale.

used in conjunction with the soil column modeling results to conduct sensitivity analysis. Shown in Figure 4 are the concentrations of Co(II)EDTA in the fracture and matrix. Because the movement of Co(II)EDTA in the matrix is relatively slower than that in the fracture, the majority of Co(II)EDTA in the matrix, except the area near the source well (red oval near the left boundary), is a result of physical mass transfer. Nonetheless, the distribution of Co(II)EDTA between the aqueous and solid phases in the fracture domain (Figure 5) suggests that, at field scale, chemical mass transfer may have a more profound impact relative to physical mass transfer on the transport of Co(II)EDTA. Note that both figures show calculations at approximately 20 hours since the release of Co(II)EDTA was started.

Co(II)EDTA, once absorbed on solid surface, undergoes a quick transformation, mediated by MnO_2, to Co(III)EDTA. As a result, surface MnO_2 is quickly consumed near the source area. Downstream from the source area, aqueous free Mn and adsorbed Mn concentration steadily increase as upstream MnO_2 is dissolved (data not shown). Note that the spatial distribution of surface MnO_2 is assumed uniform initially over the entire transect. As of the writing of the paper, effort is underway to incorporate chemical heterogeneity in the WAG 5 model and to study the effect of spatial variation in MnO_2 on the fate and transport of Co(II)EDTA.

5 ACKNOWLEDGEMENT

The authors are greatly indebt to the financial support of Dr. Paul Bayer in the DOE Environmental Technology Partnership Program.

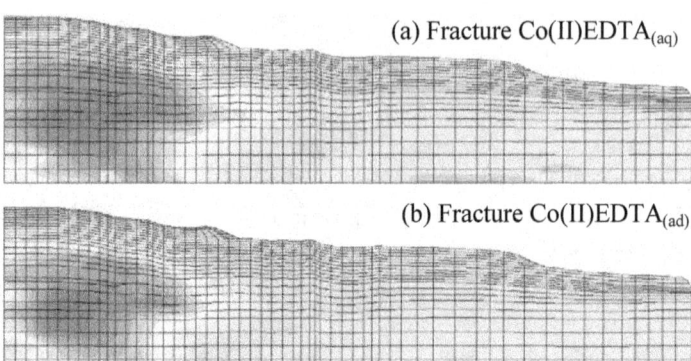

(a) Fracture Co(II)EDTA(aq)

(b) Fracture Co(II)EDTA(ad)

Figure 5. Predicted Co(II)EDTA concentrations in the (a) aqueous and (b) solid phases in the fracture. Note that concentrations are plotted in logarithmic scale.

REFERENCES

Allen, D. M., and F. A. Michel, 1999, Characterizing a faulted aquifer by field testing and numerical simulation, *Ground Water*, 37: 718-728.

Bodvarsson, G.S., Yu-Shu Wu, and Keni Zhang, 2003, Development of discrete flow paths in unsaturated fractures at Yucca Mountain, *J. Cont. Hydrol.* 62-63: 23 - 42.

Gerke, H.H., and M. T. van Genuchten,1993, A dual-porosity model for simulating the preferential movement of water and solutes in structured porous media, *Water Resour. Res.* 29: 305 - 319.

Gwo, J.P., Jardine, P.M., Yeh, G.T., Wilson, G.V., 1994. *MURF User's Guide: A Finite Element Model of Multiple-Pore-Region Flow Through Variably Saturated Subsurface Media*, ORNL/GWPO-011, Oak Ridge Natl. Lab. Oak Ridge, Tenn.

Gwo, J. P., P. M. Jardine, G. V. Wilson, and G. T. Yeh, 1996, Using a multiregion model to study the effects of advective and diffusive mass transfer on local physical nonequilibrium and solute mobility in a structured soil, *Water Resour. Res.* 32: 561-570.

Gwo, J.P., Jardine, P.M., Yeh, G.T., Wilson, G.V., 1995. *MURT User's Guide: A Hybrid Lagrangian-Eulerian Finite Element Model Of Multiple-Pore-Region Solute Transport Through Subsurface Media*. ORNL/GWPO-015, Oak Ridge National Laboratory.

Gwo, J. P., P. M. Jardine and W.E. Sanford, 2004a. Effect of advective mass transfer on field scale fluid and solute movement: field and modeling studies at a fractured waste disposal site at Oak Ridge National Laboratory, Tennessee, Hydrogeol. J. (in press).

Gwo, J. P., P. M. Jardine and W.E. Sanford, 2004b. Uncertainty in interpretation of field-scale multiple tracer injection data at a fractured low-level waste disposal site, *J. Cont. Hydrol.* (in review).

James, B. R., J. P. Gwo, and L. E. Toran. 1996. Risk-cost decision framework for aquifer remediation design. *J. Water Resour. Plan. and Manag.* 122: 414-420.

Jardine, P. M., Sandford W. E., Gwo J. P., Reedy, O. C., Hicks D. S., Riggs J. S., Bailey W. B., 1999. Quantifying diffusive mass transfer in fractured shale bedrock, *Water Resour. Res.* 35: 2015 – 2030.

Jardine, P. M., T. L. Mehlhorn, I. L. Larsen, W. B. Bailey, S. C. Brooks, Y. Roh, and J. P. Gwo, 2002.Influence of hydrological and geochemical processes on reactive contaminant transport in fractured shale bedrock, *J. Cont. Hydrol.* 55(2002): 137-159.

Lee, R.R, R.H. Ketelle, J.M. Bownds, T.A. Rizk, 1992, Aquifer analysis and modeling in a fractured, heterogeneous medium, *Ground Water*, 30: 589 - 597.

Neretnieks I (1980) Diffusion in the rock matrix: An important factor in radionuclide retardation?, *Water Resour. Res.* 85: 4379 – 4397.

Novakowski KS, Lapcevic PA (1994) Field measurement of radial solute transport in frractured rock, *Water Resour. Res.* 30: 37-44.

Sudicky, E. A., and R. G. McLaren, 1992, The Laplace transform Galerkin technique for large-scale simulation of mass transport in discretely fractured porous formations, *Water Resour. Res.* 28, 499-512.

Tsang, Y. W., and C. F. Tsang, 1987, Channel model of flow through fractured media, *Water Resour. Res.* 23: 467-479.

4.12 Radionuclide Sorption in High-Level Waste Performance Assessment: Abstraction of Results from Experiments and Surface-Complexation Models

R.T. Pabalan[1], D.R. Turner[1] and F.P. Bertetti[1]
[1]Center for Nuclear Waste Regulatory Analyses, Southwest Research Institute®, San Antonio, Texas (rpabalan@swri.org)

Sorption onto minerals present along groundwater flow paths may help to mitigate radionuclide transport from the potential high-level nuclear waste repository at Yucca Mountain, Nevada. Radionuclide transport models used in probabilistic performance assessment calculations typically use a constant sorption coefficient (K_d) for each radionuclide and each hydrostratigraphic unit. However, constant K_d models typically do not account for spatial variability in the mineralogy of the adsorbing phases or for variable chemical conditions (i.e., pH, ionic strength, alkalinity, and concentrations of complexing ligands) that may be encountered along groundwater flow paths. In this presentation, alternative approaches are discussed that incorporate aspects of mechanistic sorption models into probabilistic performance assessment calculations. A diffuse-layer surface complexation model is calibrated against laboratory experiments and used to calculate actinide transport parameters. In one approach, parameter distributions are calculated using the diffuse-layer surface complexation model and data on the chemistry of groundwaters from the Yucca Mountain vicinity. Model results are used to provide limits on K_d probability distribution functions as input into performance assessment. Under the groundwater chemical conditions observed in the vicinity of Yucca Mountain, calculated K_ds for some actinides could range over many orders of magnitude. Another approach uses the diffuse-layer surface complexation model to calculate actinide sorption behavior for a wide range of geochemical parameters and to develop response surfaces for actinide sorption (e.g., as functions of pH and pCO_2) for use in performance assessment calculations. These alternative approaches can be adapted readily to current performance assessment abstractions using site-specific information to provide geochemical constraints on radionuclide transport parameters.

4.13 Application of Surface Complexation Modeling to Describe Uranium(VI) Adsorption and Retardation at the Field Scale

G.P. Curtis[1], J.A. Davis[1] and M. Kohler[2]
[1]USGS, Menlo Park, CA, gpcurtis@usgs.gov
[2]Colorado School of Mines, Golden, CO

1 INTRODUCTION

Uranium (U) is a contaminant of concern to the U. S. Federal Government at many sites where it has been released to the environment or where it remains in poorly designed storage facilities. At the U. S. Department of Energy facilities, U is the most common radionuclide contaminant in groundwater/sediment systems. Risk assessments must be conducted for many of these contaminated sites to evaluate remediation scenarios, and a significant component of such risk assessments includes predictions of U transport to drinking water supplies or biological receptors via a groundwater pathway.

In oxic environments, the most stable valence of uranium is U(VI) which forms moderately soluble solid phases. Thus, at concentrations less than approximately $30\mu M$, the mobility of U(VI) can be controlled by adsorption reactions at circumneutral pH values. In performance assessment model simulations, adsorption of U(VI) has been approximated by a single distribution coefficient, K_D, or a range of spatially and temporally uniform K_D values. If there is significant chemical variability, however, the K_D value may not be constant and therefore can introduce considerable uncertainty into the model simulations. This is particularly true for U(VI) which undergoes complex aqueous and surface speciation reactions. Alternatively, surface complexation models (SCMs) can be used to describe U(VI) adsorption for variable geochemical conditions. In this study, a semi-empirical SCM was calibrated to laboratory data for adsorption of U(VI) by aquifer sediments and then this SCM was tested in field studies and in reactive transport simulations.

2 SITE DESCRIPTION

A uranium mill that was located along the San Miguel River near the town of Naturita, CO processed uranium ores by salt roasting followed by both sodium carbonate and sulfuric acid extractions to produce uranium concentrates between 1939 and 1961. Processed mill tailings were removed from the site in 1979 and contaminated soils were removed from the site between 1996 and 1998. Figure 1 shows the locations of all of the active groundwater wells that were installed in the shallow, unconfined alluvial aquifer that consists of cobbles, gravels, sands and some fine material. The aqueous U(VI) concentrations observed in June 2000 are also shown. U(VI) varied from background concentrations of 0.02

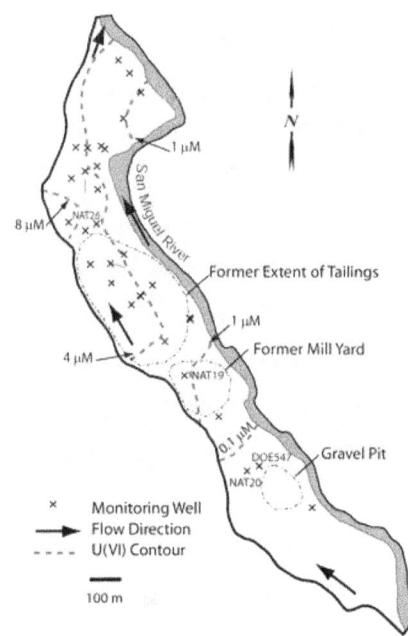

Figure 1. Naturita mill tailings site illustrating the monitoring wells, U(VI) concentrations, the extent of the former tailings piles and mill site, and the gravel

μM at DOE547 to maximum concentrations of 10.2 μM at NAT26. The spatial distribution of alkalinity was similar in shape to that for dissolved U(VI) but the values ranged from 4 meq/L in background wells to 12 meq/L at NAT26. The pH values were typically between 6.8 and 7.2 and did not illustrate a discernable, persistent spatial pattern. The waters were generally close to saturation with calcite, which is present in the sediments.

3 APPROACHES TO ADSORPTION MODELING

Uranium(VI) adsorption onto Naturita aquifer background sediments (NABS) was studied in batch experiments as a function of pH, U(VI) and CO_2 partial pressure in artificial groundwater solutions. The sediments were collected from the alluvial aquifer at a gravel pit located upgradient from the former uranium mill (Figure 1). The ranges of aqueous chemical conditions used in the U(VI) adsorption experiments (pH 6.9 to 7.9; U(VI) concentration 0.02 to 10 μM; partial pressure of carbon dioxide gas 0.05 to 6.8%) were based on chemical conditions observed in 1999-2000 in the Naturita alluvial aquifer. The major minerals in the sediments were quartz, feldspars, and calcite, with minor amounts of magnetite and clay minerals. Quartz grains commonly had coatings that were greater than 10 nm thick and composed of an illite-smectite clay with occluded ferrihydrite and goethite particles.

A semi-empirical SCM was calibrated to laboratory adsorption data that used Naturita sediments and the observed ranges in geochemical conditions. In the modeling approach, adsorption was postulated to occur on generic surface sites that represent average properties of the surface rather than specific mineral surfaces. Model parameters were calibrated to adsorption data for postulated reaction stoichiometries and different model formulations are selected on the basis of simplicity and goodness of fit. Electrostatic terms were not included in the model because of the difficulties in quantifying the electrical field and charge at the mineral-water interface in the mixture of mineral phases and associated surface coatings. Figure 2 shows that the calibrated model reproduces measured U(VI) K_D values over the entire range of experimental (and field) conditions. The semi-empirical SCM approach is therefore a practical approach that allows the modeler to couple the effects of variable geochemical conditions on adsorption and retardation.

Published models for U(VI) adsorption on reference minerals were compared to the semi-empirical SCM in terms of the ability to describe the U(VI) experimental data based on assumptions about the sediment surface composition, surface area and electrical double layer properties. Predictions from these models were highly variable, with results over predicting or under predicting the experimental data, depending on the assumptions used to apply the model. Although the models for reference minerals are supported by detailed experimental studies (and in ideal cases, surface spectroscopy), the results suggest that errors in applying the models directly to the Naturita sediment result from uncertain knowledge of: (1) the proportion and types of surface functional groups available for adsorption in the inorganic surface coatings, (2) the electric field at the mineral-water interface, and (3) surface reactions of major ions in the aqueous phase, such as Ca^{2+}, Mg^{2+}, HCO_3^-, SO_4^{2-}, H_4SiO_4, and

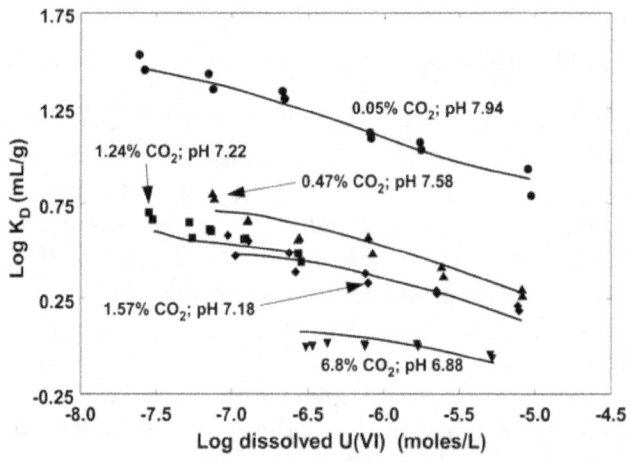

Figure 2. Fit of the semi empirical SCM to observed U(VI) adsorption data.

natural organic matter.

4 FIELD TESTS OF THE SCM

The SCM developed in laboratory studies was tested for its ability to describe U(VI) adsorption in the field using two different experimental approaches. In the first case, U(VI) adsorption onto contaminated sediments that were collected during well installation was measured using both isotopic exchange experiments and extractions with a Na_2CO_3 solution that had a pH of 9.5 and an alkalinity of 20 meq/L. The two experimental methods gave nearly identical results. In the second case, U(VI) adsorbed by NABS samples that were suspended in wells that had contaminated groundwater was determined by extraction with the Na_2CO_3 solution. Observed K_D values were calculated from adsorbed U(VI) measured in the extractions and the dissolved U(VI) measured in the wells. The observed K_D values are compared with model-predicted K_D values in Figure 3. The SCM gives model-predicted K_D values that agreed with measured K_D values within a factor of 2.2 on average. The contaminated sediments from MAU03 and MAU04 had measured K_D values that were significantly larger than the model-predicted K_D values and it is possible that these samples could have contained U(IV). The possible presence of U(IV) is consistent with the occurrence of Fe(II) in these wells; other chemical conditions at these wells are not likely explanations because measured K_D values for NABS samples in these wells agreed closely with model predictions. Overall, the agreement between model-calculated and measured K_D values increases the confidence in the applicability of the laboratory determined SCM to

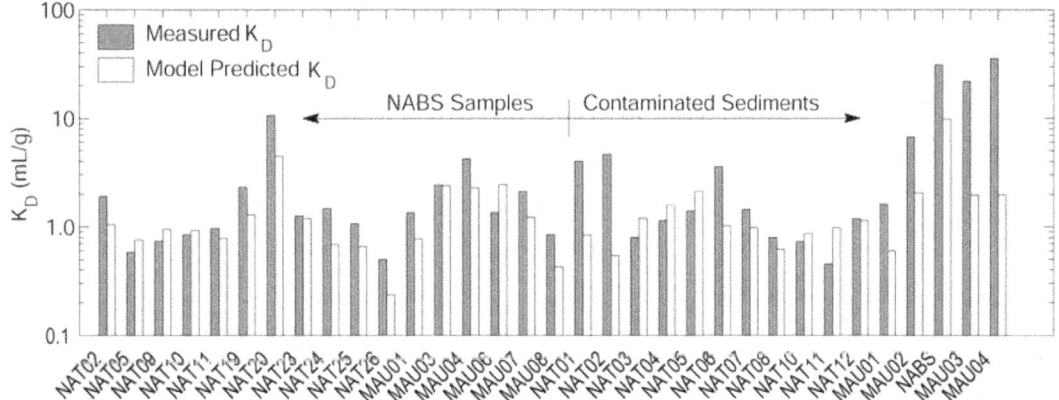

Figure 3. Comparison of measured and model-predicted K_D values for (a) uncontaminated NABS samples that were suspended in ground water wells and (b) contaminated sediments collected during well installation.

simulate U(VI) adsorption in the field.

5 REACTIVE TRANSPORT MODELING

A groundwater flow and transport model was developed for the site using tritium-helium age dating results and a Cl plume that originated from residues from the salt roasting of the U ore. A reactive transport model was calibrated to the existing site data by varying the U(VI) and alkalinity concentrations in the recharge water to give a best fit to the observed values. The contaminated recharge was assumed to extend from the upgradient boundary of the former mill yard to the downstream boundary of the former tailings pile. The contaminated recharge was assumed to be present until the contaminated soils were removed from the site in 1998. The calibration of the reactive transport model did not vary the SCM parameters or the hydraulic conductivity (K_H), which were determined independently.

Figure 4 shows the simulated and observed distributions of U(VI) and alkalinity. The simulated U(VI) and alkalinity plumes generally match the observed spatial distribution in that the locations and values of the highest concentrations are similar and there are lower observed and simulated values near the river.

The log K_D values shown in Figure 5 were calculated from the simulated adsorbed and dissolved U(VI) concentrations. The simulated K_D values vary by approximately a factor of 10 and the simulated values are not distributed randomly but have significant spatial structure. The groundwater with the largest dissolved U(VI) and alkalinity is transported with the smallest K_D values. Therefore the spatial distribution of K_D values is also expected to vary temporally.

A simple sensitivity analysis was conducted to evaluate the influence of hydraulic conductivity (K_H) and SCM model parameters on the simulated U(VI) concentration. The model was most sensitive to K_H. The sum of sensitivity of the model to the SCM site concentrations was about 40% of the K_H value.

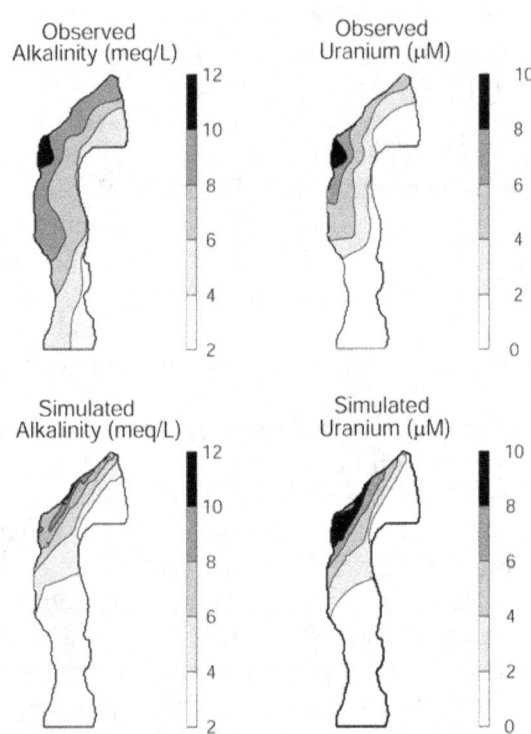

Figure 4. Comparison of observed and simulated distributions of dissolved U(VI) concentration and alkalinity in 2000.

6 SUMMARY AND CONCLUSIONS

The reactive transport model that includes the semi-empirical SCM is a useful tool for simulating the transport of U(VI) at the field scale with chemically variable conditions. This modeling approach is a compromise modeling approach between the simple constant K_D approach and the most complex of SCM modeling approaches that attempt to account for sorption on each of the many minerals present in most sediments and possibly the electrical double layer properties of each mineral phase. The semi-empirical SCM approach is based on the premise that coatings dominate the surface chemistry of most of the mineral grains, and as a consequence, adsorption can be described with a relatively small set of reactions and mass action expressions. The semi-empirical SCM modeling approach can be relatively easily applied, and it does not require a burdensome program of data collection. The model is capable of accounting for the effects of variable chemical conditions on U(VI) adsorption and transport. In the Naturita case, the most important variable chemical condition was the alkalinity, but at other sites, pH or the concentrations of other complexing ligands could be important.

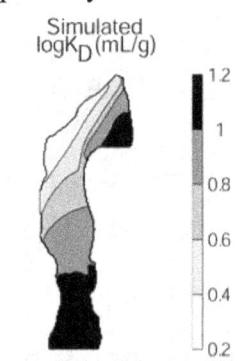

Figure 5. Simulated Kd values for the year 2000.

4.14 Reactive transport modeling of multicomponent cation exchange at the laboratory and field scale

C.I. Steefel

Lawrence Berkeley National Laboratory, Berkeley, CA, CISteefel@lbl.gov

1 INTRODUCTION

As pointed out by Appelo (1996), multicomponent ion exchange models have not been widely used in geochemical or geochemical transport modeling of subsurface processes, despite their recognized importance in regulating soil and sediment pore water compositions. This is also somewhat surprising considering the demonstrated success of ion exchange models in describing sorption and subsurface transport (Ritchie, 1966; Pope et al., 1978; Valocchi et al., 1981a and 1981b; Griffioen, 1993; Appelo and Willemsen, 1987; Appelo, 1994; Lichtner et al., 2004; Steefel, 2004). The key strength of the ion exchange models over "simpler" formulations like the linear distribution coefficient (K_d), the Langmuir isotherm, or even the Freundlich isotherm, is that it captures competitive adsorption behavior through the use of a mass action expression. The total exchange capacity, which can not be exceeded in classical ion exchange, is incorporated as well. Since the ion exchange process involves electrically balanced exchange, complexities associated with the development of surface charge such as appear in the case of electrostatic surface complexation models are avoided. In some cases, it has been necessary to introduce multi-site models to account for the strong dependence of exchange on cation ion concentration (Brouwer et al., 1983; Comans et al., 1991; Poinssot et al., 1999; Zachara et al., 2002; Steefel et al., 2003). Other difficulties have arisen in heterovalent systems where the non-ideality of exchange has required the use of activity coefficients for exchangers (Sposito, 1981; Appelo, 1996; Liu et al., 2004).

2 MATHEMATICAL FORMULATION OF CATION EXCHANGE

An ion exchange reaction can be described via a mass action expression with an associated equilibrium constant (Vanselow, 1932; Sposito, 1981; Appelo and Postma, 1993). The exchange reaction can be written in generic form as

$$vACl_u(aq) + uBX_v(s) \leftrightarrow uBCl_v(aq) + vAX_u(s) \tag{1}$$

where X refers to the exchange site occupied by the cations A^{u+} and B^{v+}. The equilibrium constant, K_{eq}, for this reaction can be written as (Vanselow, 1932)

$$K_{eq} = \frac{(BCl_v)^u (AX_u)^v}{(ACl_u)^v (BX_v)^u} \tag{2}$$

where the parentheses () refer to the thermodynamic activities. Activity corrections for solutes can be described with various models (e.g., Debye-Huckel, Pitzer), but there has been less agreement on models for calculating activity corrections for exchangers. Several activity conventions are in wide use. Vanselow (1932) suggested the use of mole fractions to represent exchange species activities, which for our generic binary system are given by

$$N_A = \frac{\{AX_u\}}{\{AX_u\} + \{BX_v\}} \qquad N_B = \frac{\{BX_v\}}{\{AX_u\} + \{BX_v\}}, \tag{3}$$

where the brackets {} refer to the concentration in moles. As pointed out by Sposito (1981), however, the activity of a component in a mixture is equal to its mole fraction only if the mixture is ideal. If the mixture is not ideal, then the activities have to be related to their mole fractions by means of an activity coefficient. In the binary exchange reaction considered in Equations 1-3, these are given by

$$f_B = \frac{(BX_v)}{N_B} \qquad\qquad f_A = \frac{(AX_u)}{N_A}. \tag{4}$$

This implies that the empirical selectivity coefficient, K_v, is related to the equilibrium constant through the expression

$$K_v = K_{eq}\frac{f_A^{\,v}}{f_B^{\,u}}, \tag{5}$$

which, after converting to logarithms, can be written in differential form as (Argersinger et al., 1950; Sposito, 1981, Liu et al., 2004)

$$d\ln K_v = v\,d\ln f_A - u\,d\ln f_B. \tag{6}$$

This expression can be combined with the Gibbs-Duhem equation,

$$N_A d\ln(f_A N_A) + N_B d\ln(f_B N_B) = 0 \tag{7}$$

which states that changes in the activity of one component must be balanced by changes in the activity of another component as mass is conserved. This provides two equations which can be solved for the two unknowns, f_A and f_B (Sposito, 1981). As pointed out by Liu et al. (2004), however, the activity of water should be included explicitly as well

$$N_A d\ln(f_A N_A) + N_B d\ln(f_B N_B) + \frac{n_w}{\{AX_u\}+\{BX_v\}}\,d\ln a_w = 0 \tag{8}$$

where n_w is the moles of water in the exchanger and a_w is the activity of water. For a binary system, $N_A + N_B = 1$, so combining Equations 6 and 8 yields

$$u\,d\ln f_B + d\left[(1-\beta_B)\ln K_v\right] = -\ln K_v d\beta_B - uvn_w d\ln a_w, \tag{9}$$

$$v\,d\ln f_A + d\left[\beta_B \ln K_v\right] = -\ln K_v d\beta_B - uvn_w d\ln a_w, \tag{10}$$

where β_B is the equivalent fraction of cation B in the exchanger defined by

$$\beta_B = \frac{v\{BX_v\}}{u\{AX_u\}+v\{BX_v\}} = \frac{v\{BX_v\}}{CEC} \tag{11}$$

and CEC is the cation exchange capacity, here defined for only a single exchange site. The chief practical difficulty in using Equations 9 and 10 to calculate exchanger activity coefficients from selectivity coefficients (K_v) is associated with the experimental determination of the number of moles of water in the exchanger phase, although this parameter is important only where significant variations in the activity of water occur. Liu et al. (2004) presented an alternative approach to solving Equations 9 and 10 that avoids the need to determine the water content of the exchanger. Some of the apparent non-ideality of exchange can be eliminated by considering multiple exchange sites with differing equilibrium constants (Zachara et al., 2002; Steefel et al. 2003; Liu et al., 2004).

3 LABORATORY STUDIES OF CATION EXCHANGE

A large number of both batch and column (flowthrough) laboratory studies of cation exchange have been carried out (e.g., Beekman and Appelo, 1990; Cernik et al., 1994; Voegelin et al., 2000; Steefel et al., 2003). Batch experimental studies have the advantage of being relatively rapid to conduct, making it possible to span a wide range of concentrations. A possible disadvantage is that it can be difficult to displace all of the cations from the exchanger in the batch experiment. Also problematic is the dissolution of soluble mineral phases, especially carbonate, that can yield erroneous determinations of the cation exchange capacity and the exchange population. This problem can be circumvented by accounting for mineral dissolution and/or precipitation by determining changes in alkalinity or other anionic species, or by using radiolabelled cations where applicable.

Flowthrough column experiments potentially circumvent the problem of incomplete exchange by eluting multiple pore volumes through the exchanger. If the eluent is collected and analyzed continuously, one can quantify the completeness of exchange. In addition to the need to collect and analyze a large number of samples, however, care must be taken to ensure that cation breakthrough curves can be determined with sufficient precision to determine cation exchange capacities and selectivity coefficients.

3.1 *Batch versus column experiment determination of selectivity coefficients*

A limited number of studies have compared determinations of selectivity coefficients and cation exchange capacity in batch and column experiments (e.g., Cernik et al., 1994; Steefel et al., 2003). Steefel et al. (2003) compared Cs^+-Na^+ exchange in Hanford sediment batch experiments reported by Zachara et al. (2002) with exchange equilibria determined in flowthrough column experiments and found that batch-derived selectivity coefficients consistently under predicted retardation in the flowthrough column experiments. There are several possible explanations for these discrepancies. One possibility is that slow, largely irreversible uptake of Cs^+ into intralamellar or frayed edge sites on illite in the sediment may account for the observation of higher selectivity for Cs^+ in the column versus batch experiments—column experiments are typically carried over longer periods of time. Alternatively, the discrepancy may be due in part to the failure to characterize completely the exchanger composition in the batch experiments. Because column experiments involve continuous elution of many pore volumes through the sediment, they are less subject to errors resulting from the failure to displace the initial exchange species population (Steefel et al., 2003).

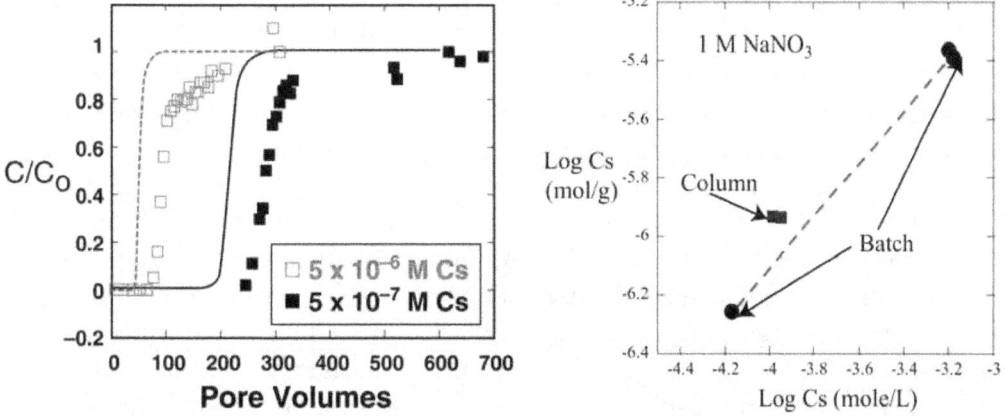

Figure 1: Left: Comparison of reactive transport simulations (lines) using selectivity coefficients determined in batch experiments with Cs^+ breakthrough determined in column experiments (symbols). Experiment conducted

in 1 M NaNO$_3$. Right: Comparison of Cs$^+$ partitioning between solid and liquid at concentrations of Cs$^+$ close to 10^{-4} M. From Steefel et al., 2003.

4   FIELD EXAMPLES OF CATION EXCHANGE

Countless examples of the application of reactive transport modeling to field examples have been presented in the literature, but in most cases these have required the calibration of modeling parameters at the field site of interest. Far rarer is the application of laboratory (batch and column) experimental data to the field. Even where laboratory determinations of selectivity coefficients and the cation exchange capacity were not made in advance, however, the ability of the multicomponent cation exchange models to capture multiple, complicated cation breakthrough curves indicates that the theory upon which the models are based is relatively robust.

Appelo (1996) and Appelo and Postma (1993) offer the most comprehensive reviews of field examples of cation exchange, along with a lucid discussion of chromatographic theory. Noteworthy examples include the case of fresh water injected into a brackish water aquifer presented by Valocchi et al. (1981a and 1981b) and discussed further by Appelo (1994, 1996) and the case of a freshening of a saline aquifer in the Aquia aquifer in Maryland (Appelo, 1994; 1996). Other field examples include those discussed in Appelo et al. (1990), Griffioen and Appelo (1993), and Beekman and Appelo (1990).

4.1   *An example of field exchange: Cs$^+$ transport in the Hanford vadose zone*

More recently, combined laboratory and field studies have been undertaken to understand the controls on Cs$^+$ transport in the vadose zone at the Hanford reservation. Cs$^+$ occurs in significant concentrations as a contaminant in highly concentrated NaNO$_3$ plumes that have developed from leaks from single-shelled storage tanks at Hanford. Zachara et al. (2002) proposed that a multi-site cation exchange model could be used to explain the Cs$^+$ behavior in the vadose zone. Steefel et al. (2003) carried out flowthrough column studies to develop a set of selectivity coefficients and exchange site concentrations that could be applied to the field-scale contamination. An important question, however, is how well these laboratory-derived parameters apply to a field setting. To test this, the case of a plume developed in the vadose zone below a leak in the SX-115 high level waste tank was used. The advantage of considering the SX-115 leak is that both its volume (about 250,000 liters) and duration (1 week in 1965) is well known. Through analysis of water extractable ions, Serne et al. (2001) identified a well-defined chromatographic separation of cations in 299-W23-19, a borehole sited about two meters from the edge of the SX-115 tank, which can be attributed to the exchange of waste-derived cations.

A direct comparison of aqueous geochemical data collected from the 299-W23-19 borehole and the 3D simulations is shown in Figure 2. The simulations capture quite closely the leading edge of all the exchange fronts, although the match with the trailing portions of the plumes is not as good. However, the simulations even capture some aspects of the complicated shape of the cation plumes, including the sodium "shoulder" above the main part of the sodium plume in the aqueous phase. This presumably represents sodium being gradually eluted from exchange sites above the main zone of sodium in the aqueous phase. The simulated widths of the magnesium, calcium, and magnesium plumes are slightly too broad, although this may be the result of an overly coarse discretization. The relatively broader shape of the potassium plume, which shows up clearly in the data, is captured in an approximate fashion by the simulations. Cesium, although not detected in significant amounts in the 299-W23-19 borehole, is known to occur very close to the bottom of the SX-115 tank (Raymond and Shdo, 1966).

It is important to point out that these results do not include any calibration of either the independently estimated hydrologic parameters (taken directly from Khaleel et al. 2001) or the laboratory-determined CEC and ion exchange selectivity coefficients. In order to match the absolute concentration of sodium in the borehole, the composition of the tank leak fluid had to be adjusted, however, with a "best fit" value representing a dilution of the estimate provided by Lichtner and Felmy (2003) by about a factor of 3.5. The exchanger calcium concentration required almost no adjustment from the value determined experimentally, while the exchange concentration of magnesium determined in the column experiments had to be adjusted downward by about a factor of 2 to obtain a match with the field data. Potassium had to be increased by a factor of 2 to 4 over the values estimated from the column experiments.

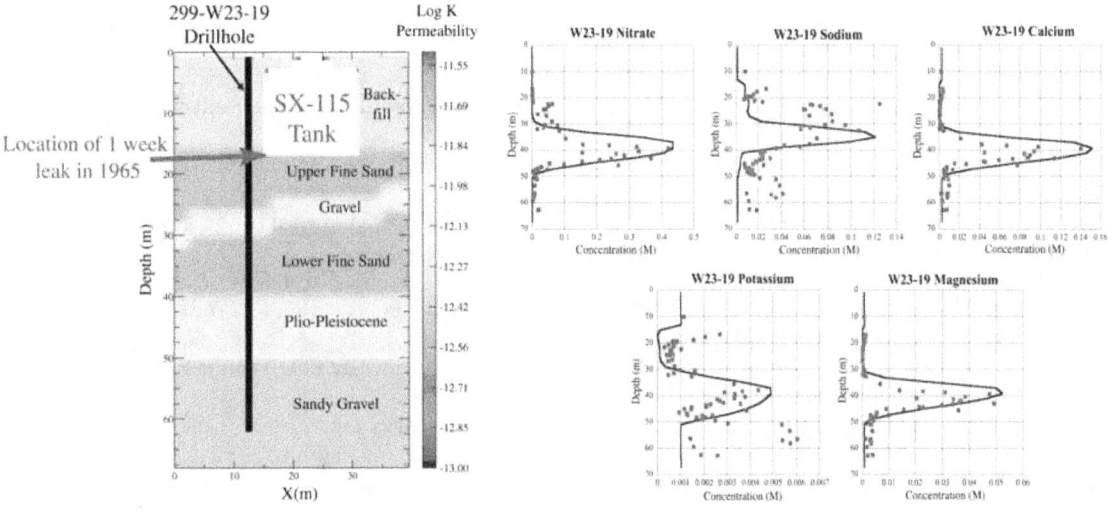

Figure 2: Left: Two-dimensional slice through 3D permeability field at SX-115 tank. Right: Comparison of 3D modeling results with pore water extracts from Borehole 299-W23-19 below the SX-115 tank at the Hanford Reservation, USA.

5 SUMMARY

Multicomponent ion exchange models have been successful in describing the chromatographic separation of cations in both laboratory and field settings. Their chief advantage lies in their ability to capture the competitive effects of other cations that may be present. By incorporating exchanger activity coefficients calculated on the basis of the Gibbs-Duhem equation applied to the exchanger phase, it is possible to correct for the non-ideality of exchange. The use of multiple exchange sites can also substantially improve the ability of the cation exchange models to describe adsorption and retardation. All of these benefits are associated with relatively little additional computational burden. Even where the cost of the multicomponent cation exchange calculations are considered too high, the models are useful in calculating distribution coefficients for the environmental conditions of interest.

REFERENCES

Appelo, C.A.J., 1994. Some calculations on multicomponent transport with cation-exchange in aquifers. *Ground Water* 32: 968-975.

Appelo, C.A.J. 1996, Multicomponent ion exchange and chromatography in natural systems. In: P.C. Lichtner, C.I. Steefel, and E.H. Oelkers (Editors), *Reactive Transport in Porous Media*. Reviews in Mineralogy 34: 193-227.

Appelo, C.A.J. and Postma, D. 1993, *Geochemistry, Groundwater, and Pollution*. A.A. Balkema, Rotterdam.

Appelo, C.A.J. and Willemsen, A. 1987. Geochemical calculations and observations on salt water intrusions, I. *J Hydrology* 94: 313-330.

Appelo, C.A.J., Willemsen, A., Beekman, H.E. and Griffioen, J. 1990. Geochemical calculations and observations on salt water intrusions, II. *J Hydrology* 120: 225-250.

Argersinger, W.J., Davidson, A.W., and Bonner, O.D. 1950. Thermodynamics and ion exchange phenomena. *Trans. Kansas Acad. Sci.* 53: 404-410.

Beekman, H.E. and Appelo, C.A.J. 1990. Ion chromatography of fresh- and salt-water displacement: laboratory experiments and multicomponent transport modeling. *J. Contam. Hydrology* 7: 21-37.

Brouwer E., Baeyens B., Maes A., and Cremers A. 1983. Cesium and rubidium ion equilibria in illite clay. *J. Phys. Chem.*, 87: 1213-1219.

Cernik, M., Barmettler, K., Grolimund, D., Rohr, W., Borkovec, M., Sticher, H. 1994. Cation transport in natural porous media on laboratory scale: multicomponent effects. *J. Contam. Hydrology* 16: 319-337.

Comans, R.N., Haller, J.,M. and DePreter, P. 1991. Sorption of cesium on illite: Non-equilibrium behavior and reversibility. *Geochim. Cosmochim. Acta* 55: 433-440.

Griffioen, J. 1993. Multicomponent cation exchange including alkalinization/acidification following flow through sandy sediment. *Wat. Resources Res.* 29: 3005-3019.

Griffioen, J. and Appelo, C.A.J. 1993. Adsorption of Ca^{2+} and CaHCO3+ by exchangers in Ca-H-Cl-CO2 systems. *Soil Sci. Soc. Am. J.* 57: 716-722.

Khaleel, R., T. E. Jones, A. J. Knepp, F. M. Mann, D. A. Myers, P. M. Rogers, R. J. Serne, and M. I. Wood 2000. Modeling Data Package for S-SX Field Investigation Report (FIR), *RPP-6296, Rev. 0*, CH2M Hill Hanford Group, Inc., Richland, WA.

Lichtner, P.C. and Felmy, A.R. 2003. Estimation of Hanford SX tank waste compositions from historically derived inventories. *Computers & Geosciences* 29: 371-383.

Lichtner, P.C., Yabusaki, S., Pruess, K., and Steefel, C.I. 2004. Role of competitive cation exchange on chromatographic displacement of cesium in the vadose zone beneath the Hanford S/SX tank farm. *Vadose Zone J.* 3: 203-219.

Liu, C. Zachara, J.M., and Smith, S.C. 2004. A cation exchange model to describe Cs^+ sorption at high ionic strength in subsurface sediments at Hanford site, USA. *J. Contam. Hydrol.* 68: 217-238.

Poinssot C., Baeyens B., and Bradbury M.H., 1999. Experimental and modeling studies of cesium sorption on illite. *Geochim. Cosmochim. Acta*, 63: 3217-3227.

Pope, G.A., Lake, L.W., and Helfferich, F.G. 1978. Cation exchangein chemical flooding: Part 1—Basic theory without dispersion. *Soc. Petrol Eng. J.* 418-433.

Raymond, J. R., and Shdo, E. D. 1966, Characterization of subsurface contamination in the SX tank farm, *BNWl-CC-701*, Richland, WA.

Ritchie, A.S. 1966. Chromatography as a natural process in geology. In Giddings J.C., Keller, R.A. (eds.) *Advances in Chromatography* 3: 119-134.

Serne, R.J., Schaef, H.T., Bjornstad, B.N., Lanigan, D.C., Gee, G.W., Lindenmeier, C.W., Clayton, R.E., LeGore, V.L., O'Hara, M.J., Brown, C.F., Orr, R.D., Last, G.V., Kutnyakov, I.V., Burke, D.B., Wilson, T.C. and Williams, B.A., 2001a. Geologic and geochemical data collected from vadose zone sediments from borehole 299 W23-19 [SX -115] in the S/SX Waste Management Area and Preliminary Interpretations. PNNL-2001-3, Pacific Northwest National Laboratory, Richland, Washington.

Sposito, G. 1981. *The Thermodynamics of Soil Solutions*. Oxford, Clarendon Press, New York, USA.

Steefel, C.I. 2004. Evaluation of the field-scale cation exchange capacity of Hanford sediments. *2004 Water-Rock Interaction Conference*.

Steefel, C.I., Carroll, S.A., Zhao, P. and Roberts, S. 2003. Cesium migration in Hanford sediment: a multisite cation exchange model based on laboratory transport experiments. *J. Contam. Hydrol.* 67: 219-246.

Valocchi, A.J., Roberts, P.V., Parks, G.A. and Street, R.L. 1981a. Simulation of the transport of ion-exchanging solutes using laboratory-determined chemical parameter values. *Ground Water* 19: 600-607.

Valocchi, A.J., Street, R.L. and Roberts, P.V. 1981b. Transport of ion-exchanging solutes in groundwater: Chromatographic theory and field simulation. *Wat. Resources Res.* 17: 1517-1527.

Vanselow, A.P. 1932. Equilibria of the base-exchange reactions of bentonites, permutites, soil colloids, and zeolites. *Soil Sci.* 33: 95-113.

Voegelin, A., Vulava, V.M., Kuhnen, F. and Kretschmar, R. 2000. Multicomponent transport of major cations predicted from binary adsorption experiments. *J. Contam. Hydrol.* 46: 319-338.

Zachara, J.M., Smith, S.C., Liu, C., McKinley, J.P., Serne, R.J. and Gassman, P.L., 2002. Sorption of Cs^+ to micaceous subsurface sediments from the Hanford site, USA. *Geochim. Cosmochim. Acta* 66: 193-211.

4.15 Conceptual Approaches for Scaling from Molecular to Macroscopic Levels of Nucleation and Precipitation

N. Sahai

Department of Geology and Geophysics, University of Wisconsin, Madison, 53706, sahai@geology.wisc.edu.

1 INTRODUCTION

1.1 *Motivation*

Precipitation and dissolution, by destroying and creating void space, can profoundly alter the fabric of rock, waste-containment cannisters, or reactive barriers, thus, impacting mass transfer and fluid flow dynamics in the porous medium. The aims of this study are to determine the molecular-level controls on mineral precipitation, and to examine conceptual approaches for scaling up to the macroscopic level.

1.2 *The Problem*

Precipitation may be viewed as two sub-processes, nucleation and crystal growth. Both are path-dependent, and affect precipitation rate. Nucleation, in particular, is poorly understood. At low temperatures, aqueous nucleation can be extremely slow, placing practical limitations on the observation time-scale. Heterogeneous nucleation reduces the time-scale, but added complexities arise such as identification of the type and number of active surface sites. A third problem relates to spatial scales. The earliest formed solid phases are nanometer-sized. Their characterization becomes problematic, because most spectroscopic methods yield atomic level information (Å scale), while traditional microscopic methods work, of course, in the micron size-range. Furthermore, reaction mechanisms are chemically complex. The earliest precipitated phases are often not thermodynamically the most stable (Ostwald Rule of Stages). Worse, thermodynamic stability, itself, is a function of particle size. Precursor phases transform to the most stable form over time. The sequence of aqueous oligomers and solid precursors determines the reaction pathway and, thus, the rate of nucleation and crystal growth. Minor amounts of other dissolved species can also significantly affect the pathway. Finally, even if the chemical complexity is fully appreciated and understood, it is difficult to incorporate the information into a useful form such as a reactive flow model.

1.3 *The Approach*

A two-pronged tack is presented here for further discussion and debate. First, I link results of molecular orbital (MO) calculations of energy and vibrational frequencies, for heterogenous apatite nucleation, to vibrational spectroscopy results. I then suggest links between molecular modeling approaches and microscopy methods. Finally, I review phenomenological models (PM) to nucleation. Values of PM parameters estimated by molecular modeling and spectroscopy/microscopy, can ultimately be incorporated into reactive flow models.

2.1 *Calcium Phosphate Precipitation*

Computational details are provided elsewhere (Sahai and Tossell, 2000; Sahai, 2003). Stable intermediates involved in nucleation were modeled, in consecutive elementary steps, as $Ca(H_2O)_6^{2+}$ adsorption at different surface sites followed by $HPO_4(H_2O)_4^{2-}$ adsorption. The effects of solvation, Mg^{2+} and H^+ were also considered. Realistic constraints were provided by comparing predicted vibrational spectra to experimental spectra.

The silicon three-ring ($Si_3O_6H_6$) was identified as the active site. The most likely reaction pathway involves partial dehydration of the calcium ion leading to inner-sphere adsorption, followed by attachment of HPO_4^{2-}, resulting in the formation of a $CaHPO_4$ critical nucleus. In terms of model clusters, at circum-neutral pH, the pathway is represented as: $Si_3O_6H_5(H_2O)_3^- \rightarrow Si_3O_6H_5Ca(H_2O)_6^+ \rightarrow Si_3O_6H_5CaHPO_4(H_2O)^-$ (Fig. 1a-c).

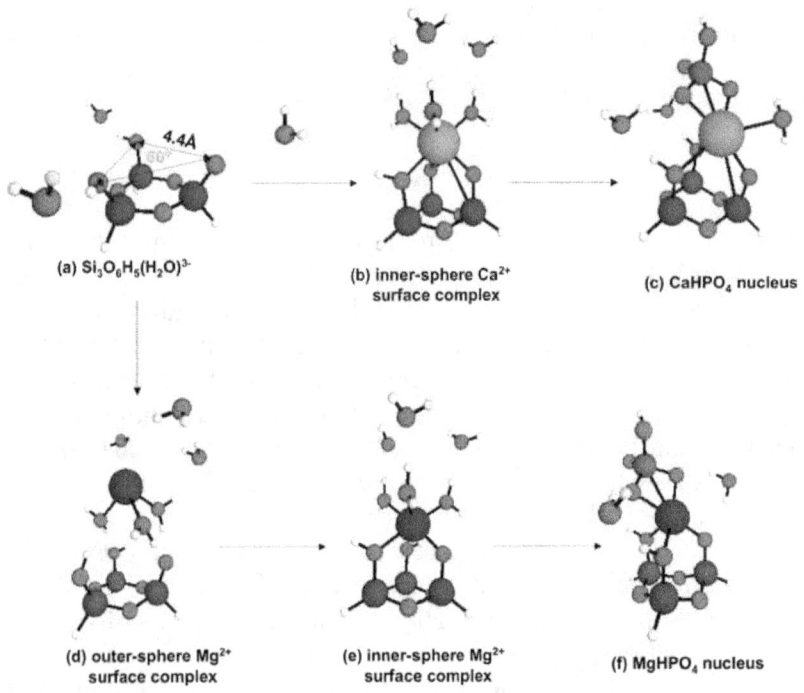

Figure 5. Stable intermediates in amorphous calcium phosphate (a-c) and magnesium phosphate (d-f) nucleation. Heavier arrows for indicate faster elementary steps for one cation than for analogous steps involving the other cation. Legend for atoms: white, H; red, O; dark blue, Si; light grey, P; dark grey, Mg; light blue, Ca.

The electron density associated with surface silanol oxygens, and the geometry of the silanols in the three-ring promote calcium ion dehydration and, thus, formation of the $CaHPO_4$ nucleus with bond angles and bond-lengths similar to those in apatite and brushite. Significantly, $Si_3O_6H_5CaHPO_4(H_2O)^-$ and silica reacting with supersaturated solution share unique Infra-Red/Raman bands at 631 and 1125-1145 cm^{-1}. These peaks are distinct from bands observed in brushite, octacalcium phosphate and crystalline apatite, suggesting that the earliest precursor is amorphous calcium phosphate. Equally important, Si four- and seven-ring surface sites, did not yield $CaHPO_4$ clusters with vibrational frequencies similar to experimental spectra.

2.2 Kinetic Effects of Mg^{2+} and H^+

Mg^{2+} and H^+, respectively, are known to inhibit and promote apatite precipitation. The relative rates of Mg^{2+} and Ca^{2+} adsorption were predicted as Mg^{2+} outer-sphere > Mg^{2+} inner-sphere > Ca^{2+} inner-sphere (Sahai, 2003). It is noteworthy that Mg^{2+} sorption is faster than Ca^{2+}, despite dehydration for $Mg(H_2O)_6^{2+}$ being slower than for $Ca(H_2O)_6^{2+}$. The faster adsorption is driven by stronger gas-phase electrostatic attraction of Mg^{2+} to the surface site. Hence, Mg^{2+} sorbs rapidly, blocking access of nucleating surface sites to Ca^{2+}. Furthermore, the outer-sphere Mg^{2+} surface complex must convert to the inner-sphere complex before HPO_4^{2-} can attach to form the corresponding critical nucleus, $Si_3O_6H_5MgHPO_4(H_2O)_3^-$ (Fig.1 a, d-f). Comparatively, HPO_4^{2-} attachment at the Ca^{2+} inner-sphere complex is faster. Apatite nucleation is, thus, retarded but not entirely prevented by Mg^{2+}. The geochemical implication is that the Mg/Ca ratio of solutions does not need to be altered significantly from "normal" compositions by *ad hoc* processes to allow authigenic apatite precipitation. The effects of Mg^{2+} on crystal growth are beyond the scope of the present paper.

Protons affect nucleation by determining protonation state of the surface site, and of the phosphate ion. The fully protonated and partially deprotonated Si three-ring were capable of nucleating appropriate $CaHPO_4$ clusters. Moreover, surface charge density (ρ) and surface potential (ψ) of the nuclei formed depend on solution pH. In turn, interfacial tension (γ) between two stable nuclei or between nucleus and solution depends on ρ and ψ according to the Lippman equation,

$$\frac{d\gamma}{d\psi} = -\rho \tag{1}$$

Interfacial tension should be minimal near the point of zero charge of the mineral, thus promoting crystal growth by expulsion of solution between two nuclei or by "particle bridging" (Schukin and Kontrovich, 1985).

2.3 Estimated Reaction Rates

Ca^{2+} and HPO_4^{2-} sorption, resulting in nucleation, occurs within minutes to $\sim$ 1 hour. Hydrolysis of the Si 3-ring occurs within $\sim$ 24 hours. Aggregation of oligomers and transformation to apatite crystals large enough to be detected by X-Ray Diffraction is the slowest step, taking about 2 weeks (Sahai and Tossell, 2000).

In summary, a combination of molecular orbital (MO) calculations with experimental spectroscopy can be used to estimate nucleation reaction mechanisms, rate determining steps, and reaction rates.

3 PHENOMENOLOGICAL MODELS (PM) OF NUCLEATION

The following discussion on nucleation borrows directly from Stumm (1992), Lasaga (1998), and Steefel and Van Cappellen (1990). Phenomenological models for crystal growth are not considered here for lack of space.

Nucleation of a mineral $A_\alpha B_\beta$ depends on the degree of supersaturation, Ω, which is related to the ion activity product (*IAP*) and macroscopic solubility (K_{so}) through:

$$\Omega = \left(\frac{IAP}{K_{so}} \right)^{1/\eta} \tag{2}$$

where, $\eta = \alpha + \beta$. For example, η for apatite written as $Ca_5(PO_4)_3(OH)$ will be different from η for $Ca_{10}(PO_4)_6(OH)_2$.

The Gibb's free energy for nucleation of a cluster containing j monomers may be represented as the sum of bulk or volume free energy, and surface free energy contributions,

$$\Delta G_j = \Delta G_{bulk} + \Delta G_{surf} = j\Delta G_r + \Delta G_{surf} \tag{3}$$

where ΔG_r is the "normal" free energy change for a macroscopic experiment.

For homogeneous nucleation,

$$\Delta G_j = -mkT \ln\left(\frac{IAP}{K_{so}}\right) + \gamma_{cw} A_{cw} \tag{4}$$

where $m = j/\eta$, is the number of formula units in the nucleating cluster, k is Boltzmann's constant, T is absolute temperature, γ_{cw} is the interfacial tension between the critical cluster (stable nucleus) and water, and A_{cw} is the surface area of the critical cluster in water.

A depends on cluster size and geometry. For example, $A = 4\pi r^2$, for a spherical cluster of radius, r. γ_{cw}, is also size dependent, so macroscopically measured values may not hold at the nanometer size range. Thus, mineral solubility depends on size. The size-dependence becomes significant at small values of r.

Critical nucleus size, or number of monomers in the critical nucleus is given by,

$$n_c^* \equiv j_c = -\frac{32\pi v^2 \gamma_{cw}^3}{3\Delta G_r^3} \tag{5}$$

where v is the volume of a monomer. For a spherical nucleus of volume, V, which is composed of j monomers, v is defined as,

$$v = \frac{4}{3\pi}\frac{r^3}{j} = \frac{V}{j} \tag{6}$$

The critical radius is,

$$r_c^* = -\frac{2v\gamma_{cw}}{\Delta G_r} = \frac{2v\gamma_{cw}}{kT\Omega} \tag{7}$$

Activation energy for nucleation is,

$$\Delta G^* = -\frac{16\pi}{3}\frac{\gamma_{cw}^3 v^2}{\Delta G_r^2} = -\frac{16\pi}{3}\frac{\gamma_{cw}^3 v^2}{(kT\ln\Omega)^2} \tag{8}$$

The rate of critical nucleus formation (nucleation rate) is given by

$$J = P\exp\left(\frac{-\Delta G^*}{kT}\right) \tag{9}$$

where the pre-exponential factor, P, is related to collision efficiency. Thus, J depends on collision efficiency, extent of supersaturation, interfacial tension, and temperature.

For heterogeneous nucleation, the surface energy contribution to ΔG_j is,

$$\Delta G_{surf} = \gamma_{cw} A_{cw} + \left(\gamma_{cs} - \gamma_{sw}\right) A_{cs} \tag{10}$$

where γ_{cs} is the interfacial tension between the critical nucleus and the substrate surface, and A_{cs} is the surface area of the critical nucleus at the substrate surface. It is, thus, the difference in interfacial tensions between the nucleus at the substrate and the nucleus in solution that drives heterogeneous nucleation. If the geometry of the substrate surface site matches the nucleating crystal, then, $\gamma_{cs} < \gamma_{cw}$. For epitaxial templation, $\gamma_{cs} \prod 0$ and $\gamma_{sw} \approx \gamma_{cw}$, so the surface energy term becomes,

$$\Delta G_{surf} = \gamma_{cw}\left(A_{cw} - A_{cs} \right) \tag{11}$$

The critical free energy for heterogeneous nucleation may be approximated as (Steefel and Van Cappellen, 1990),

$$\Delta G^{*} = -\frac{1\pi}{3}\frac{\gamma_{cw}^{3}v^{2}}{(kT\ln\Omega)^{2}} \tag{12}$$

The Ostwald-Lussac Rule of Stages is a kinetically driven phenomenon. From Eq. (7) above, we see that at constant Ω and v, r_c^{*} is proportional to γ. Also, it is observed empirically that for ionic solids such as carbonates, oxides and sulfates, γ_{cw} increases as solubility decreases. The most soluble phase, therefore, has the lowest interfacial tension, hence the smallest critical nucleus radius, and the largest nucleation rate. The most soluble phase, therefore, precipitates out first, and eventually transforms to the least soluble (most stable) phase. This explains why the amorphous calcium phosphate precipitates first, and later transforms to apatite.

At supersaturations above the critical value, surface area is generated by nucleation, and at Ω values less than critical, new surface area is generated by crystal growth. It is empirically observed that a polydisperse crystal suspension eventually grows to a larger, monodisperse crystal size, a recrystallization process called Ostwald Ripening. The size-dependence of solubility can explain this phenomenon. The smaller crystals are more soluble, hence redissolve, establishing a higher concentration of solutes in their vicinity. Solute concentration is depleted around larger crystals. This establishes a concentration gradient between the two crystals, which drives solute from the smaller towards the larger crystal. Thus, the larger crystal grows at the expense of the smaller ones. Thus, Ostwald Ripening is a thermodynamically driven phenomenon. Other crystal growth mechanisms include attachment of single atoms or oligomers to nuclei, aggregation of stable nuclei by particle bridging, expulsion of solution between adjacent nuclei, and epitaxial templating.

4 ESTIMATING PM PARAMETER VALUES

The discussion above suggests that a description of nucleation requires the knowledge of several model parameters such as interfacial tensions (γ_{cw}, γ_{cs}), the size-dependence of these interfacial tension terms, the extent of supersaturation (Ω), and the critical cluster size (r_c^{*}). Similarly, phenomenological models for crystal growth rates, for example, growth by attachment at step-sites, requires knowledge of parameters such as step height, spacing and velocity (e.g., Teng et al., 2000). If these parameters can be estimated or measured, a first approximation to nucleation and crystal growth becomes possible.

For instance, in a solution supersaturated with respect to apatite, the earliest detectable crystals by light scattering experiments, were 0.7-1.0 nm in size (Onuma and Ito, 1998), which we will take as an estimate of the homogenous critical nucleus size. The MO calculations yield $\sim$ 0.4- 0.45 nm for the heterogeneous $CaHPO_4$ nucleus (Fig. 1c). We infer that the heterogeneous critical nucleus is a monomer (j ~1), whereas the homogeneous critical nucleus must be a dimer or trimer of $CaHPO_4$, i.e., that $j \sim 2$-3. Further, assuming a spherical nucleus, we estimate $A_{cs} \sim 2$-2.5nm^2 and $A_{cw} \sim 6$-12 nm^2. In principle, ΔG_j values may be obtained directly from very high-level MO, MD or MC calculations (in practice, accuracy is an issue), and may be combined with the estimated A_{cs} and A_{cw}, to obtain values for γ_{cw} and γ_{cs}.

High Resolution Transmission Electron Microscopy and Scanning Probe Microscopies coupled with computational simulation methods such as Molecular Dynamics (MD) or atomistic Monte Carlo (MC) are important for a mechanistic explanation of the nanometer to micron scale observations. Equally significant, microcalorimetric methods, which determine thermodynamic properties such as heat capacity, entropy, enthalpy and Gibb's free energy of nanoparticles, are important for predicting the relative stability sequence of precursor phases (e.g., Zhang et al., 1999; Majzlan e al., 2000). Finally, a relatively new technique, Vertical Scanning Interferometry, can be coupled with MC simulations to link micron scale processes with the millimeter scale (Lasaga and Lüttge, 2001; Arvidson et al., 2003).

5 INCORPORATION OF PRECIPITATION MODELS IN REACTIVE FLOW MODELS

Mass transport by flow through porous media can be simulated using reactive flow models, which account for mass, charge, momentum, and energy balance. The mass balance equations usually contain expressions for total dissolved concentration of solutes ('basis species') distributed among different species such as the uncomplexed basis species, aqueous complexes, surface complexes, ion-exchanged complexes, and precipitated solids (e.g. Yeh and Tripathi, 1991). An additional type of species, "surface precipitates" could be included. A mass action equation with the corresponding equilibrium constant is required to determine the concentration of each species. In order to include nucleation and crystal growth, PM parameters such as γ_{cw}, γ_{cs}, r_c^*, step height, spacing and velocity, solubility constants (K_{so}) as a function of cluster radius, adsorption constants (K_{ads}) as a function of adsorbent particle-size, heat capacity, entropy and enthalpy as a function of cluster radius, type and number of reactive surface sites, etc. must be included in the thermodynamic database for the flow model. If cluster size was tracked as the reaction progresses computationally, the relative importance of the size-dependent terms would diminish over time, and it may become possible to "turn off" those terms at some critical value, and allow macroscopic properties to dominate, thus saving on computational time.

In summary, the paper presents the problems associated with modeling precipitation from the molecular level (nucleation) to macroscopic levels (bulk precipitation). Some approaches are presented for tackling these issues.

REFERENCES

Arvidson, R.S., Ertan, I.E., Amonette, J.E. and Luttge, A. 2003. Variation in calcite dissolution rates: A fundamental problem? *Geochim. Cosmochim. Acta* **67**, 1623-1634.

Lasaga, A.C. and Luttge, A. 2001. Variation of crystal dissolution rate based on a dissolution stepwave model. *Science* 291, 2400-2404.

Majzlan J., Navrotsky A. and Casey W.H. 2000. Surface enthalpy of boehmite. *Clays. Clay Min.* 48, 699-707.

Onuma, K. and Ito, A. 1998. *Chem. Mater.* 10, 3346-3351.

Sahai, N. 2003. The effects of Mg^{2+} and H^+ on apatite nucleation at silica surfaces. *Geochim. Cosmochim. Acta* 67, 1017-1030.

Sahai, N. and Tossell, J.A. 2000. Molecular orbital study of apatite ($Ca_5(PO_4)_3OH$) nucleation at silica bioceramic surfaces. *J. Phys. Chem. B.* 104, 4322-4321.

Schmidt, M.W. et al. 1993. General atomic and molecular electronic structure system *J. Comput. Chem.* 1347-1363.

Schukin, E.D. and Kontorovich, S.I. 1985. Formation of contacts between particles and development of internal stresses during hydration processes. In *Materials Science of Concrete III*, Ed. J. Skalny. The American Ceramic Society, Westerville, OH, 1992, pp. 1-35.

Steefel, C.I. and Van Cappellen, P. 1990 A new kinetic approach to modeling water-rock interaction: The role of nucleation, precursors, and Ostwald ripening. *Geochim. Cosmochim. Acta* 54, 2657-2677.

Stevens, W.J., Krauss, M., Basch, H. and Jansen, P.G. 1992. Relativistic compact effective core potentials and efficient shared exponent basis sets for the third, fourth and fifth row atoms. *Canad. J. Chem.* **70**, 612-630.

Stumm, W. 1992. *Chemistry of the Solid-Water Interface.* Wiley, NY, Chapter 6.

Teng, H.H., Dove, P.M. and De Yoreo J.D. 2000 Kinetics of calcite growth: Surfce processes and relationships to macroscopic rate laws. *Geochim. Cosmochim. Acta* 64, 2255-2266.

Yeh, G.-T. and Tripathi, V.S. 1991. A model for simulating transport of reactive multispecies components: Model development and demonstration. *Wat. Res. Res.* 27 3075-3094.

Zhang H., Penn R.L., Hamers, R.J. and Banfield, J.F. 1999. Enhanced adsorption of molecules on surfaces of nanocrystalline particles. *J. Phys. Chem. B.* 103, 4656-4662.

4.16 Determining Dissolution, Precipitation and Nucleation Rate Laws in Natural Systems

A.E. Blum
U. S. Geological Survey, Boulder, Colorado, aeblum@usgs.gov

1 INTRODUCTION

Many chemical changes in hydrologic systems are the result of dissolution and precipitation reactions. There is a growing body of theoretical work describing the dissolution and precipitation process. However, the factors potentially controlling dissolution and precipitation are very diverse, and it is difficult to design definitive experiments that will accurately reproduce the hydrologic and chemical complexity of natural systems. In addition, geologic time scales may be very important in natural systems, and are impossible to duplicate experimentally. Consequently, there is difficulty in predicting rate laws for dissolution and precipitation reactions which can be incorporated in reaction-transport modeling.

2 SPARINGLY SOLUBLE SALTS *(MOSTLY CALCITE)*

The behavior of sparingly soluble salts, the most important of which is calcite, appears to be similarly in experimental and natural systems. Thus, experimental determinations of dissolution and precipitation rates under the appropriate conditions, such as the rates of calcite precipitation determined by Plummer et al. (1978) are probably a reliable prediction of the kinetics of those processes in natural systems. Observations of the surface evolution and reaction mechanisms sparingly soluble salts seem consistent with the general theoretical approaches to crystal dissolution as growth. The details may not be clear, but the approach is insightful and significant. Commonly, the reaction rates of sparingly soluble salts are rapid enough that reaction-transport models may use the assumption of local equilibrium, and not introduce kinetics.

As always, there are notable cautions and exceptions. One of the most common examples is the severe inhibition of calcite dissolution by even trace amounts of dissolved phosphate Berner and Morse (1974). This was discovered by oceanographers trying to explain the depression of the calcite compensation depth in the oceans. Inhibiting effects have been investigated in several systems, but there are a nearly infinite combination of possible salts and inhibitors, and the possibility of inhibition must always be considered. The good news is that testing inhibition in the laboratory for a specific system is practical.

Another caveat is that composition of the solution can play a critical role in the kinetics of dissolution and precipitation reactions. For example, Plummer et al. (1992) experimentally measured the dissolution kinetics of aragonite-strontianite solid solutions [(Ca,Sr)CO$_3$]. What they found is that the composition of the solid phase had no relationship to the dissolution rate. Rather, the dissolution kinetics was controlled by the rate of dissolution of the solid which predicted to be precipitated from the solution Ca/Sr ratio. What they believe was happening is that there is reversible dissolution and precipitation occurring at the surface, consistent with the theory of microscopy reversibility (that the net reaction rate is the difference between the forward and reverse reactions at the surface on a microscopic scale). Initially the bulk crystal dissolves, but the composition of the precipitate in the back reaction is determined by composition of the solution. Within minutes, the crystal surfaces are completely covered by a thin layer reflecting the composition of the solution, and it is the solution composition which then controls the kinetics of any further dissolution.

3 DISSOLUTION OF ALUMIO-SILICATES

3.1 *Discrepancy between experimental and natural dissolution rates*

The feldspar minerals, particularly plagioclase, often dominate weathering in soils and watersheds, and have been the focus of a large proportion of both the experimental and field studies of alumino-silicate dissolution. There are difficulties and ambiguities in both laboratory and natural feldspar dissolution rates. However, it now quite clear that there is a large discrepancy between experimentally measured dissolution rates of feldspars and dissolution rates of feldspars measured in the field, with the experimental rates 10^2 to 10^5 times faster than natural rates. This data has been most recently compiled by White and Brantley (2003). It appears that experimental dissolution rates of alumino-silicates have very little relevance to natural reaction rates, and use of the experimental rates in reaction transport models of natural systems is difficult to justify. As a consequence, research has shifted over the last ten years away from experimental weathering studies to field studies designed examine the processes operating in natural systems.

There are systematic variations observed in weathering in soils. While absolute experimental dissolution rates are far too fast, the relative dissolution rates of minerals in experimental and natural systems are roughly similar, with quartz <K-feldspar <plagioclase <amphiboles <pyroxenes <olivine. Elemental fluxes from watersheds also indicate systematic variations, with weathering rates increasing with increasing precipitation and higher temperatures (White and Blum, 1995; Olivia, et al., 2003).

3.2 *Controls on feldspar weathering rates in soils*

There is a lot of speculation, but only a rough outline of reasons for retardation of natural dissolution rates is emerging. An important observation is that most weathering occurs in the top few meters of regolith. Below the B soil horizon, there is much less change in the chemistry of soil solutions, and much less evidence of primary mineral alteration or clay formation. The near-surface is also the area of greatest biological activity. However, all the proposed effects of biota tend to increase weathering rates, and no plausible biotic mechanism for the retardation of weathering rates has been established.

The most likely mechanisms retarding feldspar dissolution rates in soils are: a) the interstitial soil solutions approach equilibrium with the feldspars, which halts further dissolution until the interstitial water is mobilized, and b) the most reactive portions of the mineral grains dissolve first, and the reactivity of the minerals decreases with the extent of weathering, and therefore roughly with age. Laboratory experiments simply can not simulate the effects of geologic time on the reactivity of minerals.

There are several lines of evidence favoring the effect of the approach to equilibrium with feldspars. The effect of saturation state on feldspar dissolution has been observed experimentally, although the details of how the dissolution rate decreases as equilibrium is approached remains uncertain (Burch et al., 1993; Oelkers et al., 1994). Soil hydrology is quite complex, and episodic wetting and drying events can mobilize and concentrate the soil solutions, and channeling and capillary tension make it difficult to quantify the flow paths of water through the soil matrix. Soil solutions can also only be sampled at times of high water content, and therefore the chemistry of the pore water is difficult to characterize accurately.

If feldspars began reacting with rain water at laboratory rates, the water in a 2 μm pore would be saturated with plagioclase within hours, and all further reaction would cease, suggesting that saturation within soil pores is likely. Soils are also the zone of major clay formation, meaning that soil pore solutions must supersaturated with respect clays for some period of time. Supersaturation with kaolinite or smectite almost requires saturation with respect to the feldspars. Finally, if soil water becomes saturated with respect to feldspars in the near surface, there will be little further reaction as the water is mobilized to greater depths, which is consistent with the major alteration in soils occurring only near the surface.

The work of White and Brantley (2003) also demonstrate a correlation between a decreasing weathering rate and increasing age of the substrate. Feldspars preferentially weather at defect sites, forming dramatic etching features which permeate the grains, and some sort of preferential weathering at defects obviously occurs. However, the age of the substrate alone can not explain the concentration of weathering at the soil surface, and therefore is probably not the complete story. If high solubility material at the soil surface is removed with time, the solution would react with material at greater depths, resulting in incomplete removal of minerals from the soil surface, and the propagation of the zone of rapid weathering downward, and this pattern is not observed. However, these mechanisms are not exclusive, and both approach to equilibrium, the reactivity of the substrate, as well as other mechanisms may operate simultaneously and to different extents in different environments.

3.3 *Feldspar weathering rates in aquifers*

Zhu et al. (in prep.) determined in situ feldspar dissolution rates along 15 flow paths within the Navajo sandstone aquifer at Black Mesa, Arizona, using the inverse mass balance calculation described in Zhu (2000). Calculated plagioclase and K-feldspar dissolution rates are $10^{-16.2\pm0.4}$ mol m^{-2} sec^{-1} and $10^{-18.1\pm0.3}$ mol m^{-2} sec^{-1}, respectively. These rates are five orders of magnitude slower than laboratory measured rates, and are among the slowest rates reported in natural systems. Calculations from published water chemistry from other aquifers suggest similar feldspar reaction rates.

The ground waters from Black Mesa are all close to equilibrium with K-feldspar and plagioclase, and approach to equilibrium is the most likely cause for retarded feldspar dissolution rates in the aquifers. However, approach to equilibrium is a necessary but not a sufficient condition; there must be another coupled mechanism that prevents groundwater reaching complete equilibrium with feldspars, and allows continued reaction down gradient at a slow but nearly constant rate. They suggest that the slow precipitation of clays (in the case of Black Mesa, smectite) is the rate limiting step for the overall weathering reaction in these groundwater aquifers. Clays form primarily in the soils, but the role of clay precipitation in limiting weathering in soils and other weathering environments is still quite uncertain. However, the precipitation rate of clays is an area where there is very little data, but a process that may have an important role in the rate of overall reactions in some circumstances.

4 NUCLEATION AND CRYSTAL SIZE DISTRIBUTIONS

There is very little information on the nucleation and growth kinetics of many important minerals, particularly clays and oxides. However, crystal size distributions (CSD) record a history of the condition under which minerals nucleated and grew. Eberl et al. (1998) has observed that there are only five basic shapes of crystal size distributions observed in a wide range of geological environments. They have attempted to relate the CSD to the solution conditions during mineral formation. For example, an asymptotic distribution with many small particles is indicative of continuous nucleation and growth with supersaturation remaining sufficient to sustain nucleation. Ostwald ripening generates a very distinctive size distribution, and nucleation followed by grow with no further nucleation yields a log-normal size distribution. There are still a number of uncertainties in the growth laws that can generate and sustain CSD's, particularly log-normal distributions, but I believe that CSD's hold a large amount of information on reaction conditions during precipitation that is not currently being utilized by many geochemists.

5 CONCLUSIONS

Experimentally determined rate laws for sparing soluble salts, such as calcite, can yield useful guidance for the rates used in reaction-transport models. However, the available data all suggest that the laboratory determined rates for the dissolution of feldspars and other alumino-silicate have little relation to the rates of these reactions in natural environments. Rather, it appears that unsaturated hydrology, approach to equilibrium and possibly the precipitation kinetics of secondary phases control the rates of natural feldspar dissolution, and most likely other alumino-silicates. Both experimental and natural data on nucleation and growth rates of minerals, particularly clays, are scarce. However, the crystal size distributions of the secondary phases can yield helpful information on chemical conditions under which the minerals formed.

6 REFERENCES

Berner, R. A. and Morse, J. W. (1974) Dissolution kinetics of calcium carbonate in sea water: IV. Theory of calcite dissolution. *Amer. J. Sci.* **274**:108-134.

Burch, T. E., Nagy, K. L. and Lasaga, A. C. (1993) Free energy dependence of albite dissoltuion kinetics at 80 C and pH 8.8. *Chem. Geol.* **105**:479-506.

Eberl, D. D., Dritz, V. A. and Srodon, J (1998) Deducing growth mechanisms for minerals from the shapes of crystal size distributions. Am. J. Sci. **298**:499-533.

Oelkers, E. H., Schott, J. and Devidal, J.L. (1994) The effects of aluminum, pH and chemical affinity on the rates of alumino-silicate dissolution reactions. *Geochimica et Cosmochimica Acta,* **58**:2011-2024.

Olivia, P., Viers, J. and Dupre, B. (2003) Chemical weathering in granitic environments. *Chem. Geol.* **202**:225-256.

Plummer L. N., Busenberg, P. D., Glynn, P. D. and Blum, A. E. (1992) Dissolution of aragonite-strontianite solid solutions in nonstoichiometric $Sr(HCO_3)_2$-$Ca(HCO_3)_2$-CO_2-H_2O solutions. *Geochimica et Cosmochimica Acta,* **56**:3045-3072.

Plummer, L. N., Wigley, T. M. and Parkhurst, D. L. (1978) The kinetics of calcite dissolution in CO_2-water systems at 5 to 60 C and 0.0 to 1.0 atm CO_2. *Amer. J. Sci.* **278**:179-216.

White A. F. and Blum, A. E. (1995) Effects of climate on chemical weathering rates in watersheds. *Geochimica et Cosmochimica Acta,* **59**:1729-1747.

White A. F. and Brantley, S. L. (2003) The effect of time o the weathering of silicate minerals: why do weathering rates differ in laboratory land field? *Chem. Geol.* **202**:479-506.

Zhu, C. (2000) Estimate of recharge from radiocarbon dating of groundwater ages and numerical flow and transport modeling. *Water Resources Research,* **36**:2607-2620.

Zhu, C., Blum, A. E. and Veblen, D. (in prep.) In situ feldspar dissolution rates in saturated aquifers.

4.17 Precipitation-Front Modeling: Issues Relating to Nucleation and Metastable Precipitation in the Planned Nuclear Waste Repository at Yucca Mountain, Nevada

J.A. Apps and E.L. Sonnenthal
Lawrence Berkeley National Laboratory, Berkeley, CA, jaapps@lbl.gov.

1 INTRODUCTION

The focus of the presentation is on certain aspects concerning the kinetics of heterogeneous reactions involving the dissolution and precipitation of unstable and metastable phases under conditions departing from thermodynamic equilibrium. These aspects are particularly relevant to transient thermal-hydrological-chemical (THC) processes that will occur as a result of the emplacement of radioactive waste within the Yucca Mountain Repository. Most important of these is a phenomenon commonly observed in altering soils, sediments and rocks, where less stable minerals precipitate in preference to those that are more stable, referred to as the Ostwald Rule of Stages, or the Ostwald Step Rule.

W. Ostwald (1897) described the phenomenon characterizing his rule (as cited in Schmeltzer et al., 1998), thus:

> "...in the course of transformation of an unstable (or metastable) state into a stable one the system does not go directly to the most stable conformation (corresponding to the modification with the lowest free energy) but prefers to reach intermediate stages (corresponding to other metastable modifications) having the closest free energy to the initial state".

This phenomenon is so widespread in natural geochemical systems, particularly under hydrothermal or low temperature conditions, that few geochemical parageneses involving the subcritical aqueous phase can be described without invoking the Ostwald Rule of Stages. Commonly observed systems where this phenomenon occurs include carbonates, silica, clay minerals, iron and manganese oxides, iron sulfides and zeolites (Morse and Casey, 1988). Simulations involving natural or anthropogenically modified reactive chemical transport must therefore be consistent with field observations describable by the Ostwald Rule.

Geochemists have long been familiar with the Ostwald Rule, but, with one exception (Steefel and Van Cappellen, 1990), have not incorporated the underlying chemical principles justifying the Rule in reactive chemical transport simulations, other than through arbitrary fixes involving the suppression of the thermodynamically more stable phases, and by prohibiting the re-dissolution of minerals.

Another issue relating to mineral metastability is the contribution of interfacial free energy to the total free energy of a geochemical system. The interfacial free energy contribution is trivial for crystal sizes in excess of 1 micrometer. However, the alteration of soils and sediments entails both the dissolution of finely crystalline products of diagenesis and heterogeneous nucleation and precipitation of new phases. The latter phases are commonly microcrystalline or amorphous, with substantial contributions of surface free energy to the total Gibbs free energy of the phase. Such contributions must be taken into account when modeling the chemical evolution of such systems, as they stabilize metastable phases and can modify aqueous species concentrations by up to two orders of magnitude. This condition is especially relevant to anthropogenically driven geochemical processes involving extreme levels of supersaturation where nucleation processes are dominant.

Furthermore, by a process known as Ostwald Ripening, larger crystallites, usually possess a lower surface free energy contribution, and being more stable, destabilize smaller coexisting crystallites of the same phase, leading to a decreased crystal size distribution, and the growth of progressively fewer crystals.

2 THEORETICAL BASIS FOR THE OSTWALD RULE OF STAGES

Ever since Ostwald stated his Rule of stages in 1897, attempts have been made to derive a theoretical basis for the rule. Now, even after 100 years, an undisputed theoretical basis for the rule has not been formulated successfully, even though field and laboratory observations supporting the rule are now too numerous to mention. A general consensus is that the rule can be explained through an understanding of nucleation theory and reaction kinetics. However, there is no rigorous basis for the rule in fact, i.e., in principle, there is no reason for the rule not to be violated, and occasional exceptions to the rule have been cited. In fact, one school of thought has attempted to formulate the rule on the basis of irreversible thermodynamics.

In natural systems, manifestations of the Ostwald rule, may in fact contain components both of kinetics, irreversible thermodynamics, and equilibrium thermodynamics. (See, for example, Duffy, 1993). Because quantitative evaluations of alteration rates in complex natural systems have hardly ever been made, it is not possible at this time to comment further on the applicability of each approach, and whether or not they can be reconciled, particularly when the complexity of the problem is fully appreciated. Perhaps, the strongest argument for preferring a kinetic treatment for the Ostwald rule, is evidence for the simultaneous growth of two or more phases of the same composition. Experimental results on a zeolitic system, cited by Barrer (1982), provide substantiation, although further examples are needed to lend credibility to this position.

3 ISSUES RELATING TO PRECIPITATION KINETICS AT YUCCA MOUNTAIN

At Yucca Mountain, Nevada, the site of a planned nuclear waste repository, metastable phase assemblages resulting from the low temperature hydrolysis of vitroclastic tuffs provide ample illustration of the Ostwald Rule of Stages. The occurrence of opal-A, and opal-CT in fractures (Carlos, 1989; Carlos et al., 1991; Vaniman, 1994; Carlos et al., 1995) and in the rock matrix (Bish and Vaniman, 1985; Chipera and Bish, 1988; Duffy, 1993: Carey et al., 1997) as products of glass alteration and possibly of devitrified tuff is evidence of metastability in the subsystem SiO_2-H_2O. Metastability in this system is widespread (Kastner, 1979; Williams et al., 1985; Barrer, 1982; Parks, 1990; Chang and Yortsos, 1994). The low temperature silica polymorphs follow the sequence opal-A - opal-CT - chalcedony (a microcrystalline form of quartz) - quartz (the stable end product) at near surface ambient temperatures. The rates of silica polymorph conversion under these conditions are extremely slow, taking millions of years (Kastner, 1979; Duffy, 1993). Because of the slow rate of silica phase transformation at near earth-surface temperatures, the silica activity in solution may be controlled by the silica polymorphs, thereby stabilizing other kinetically favored metastable phases such as the clays (Abercrombie et al., 1994) or zeolites.

Zeolite parageneses in the vitroclastic deposits at Yucca Mountain also provide evidence of adherence to the Ostwald Rule of Stages, as supported by independent observations elsewhere (Dibble and Tiller, 1981; Barrer, 1982; van Santen, 1984; Morse and Casey, 1988). Smyth (1982) noted the following zeolite sequence at Yucca Mountain:

(clinoptilolite (+ mordenite) + opal-CT) $\Rightarrow$ (analcime + quartz) $\Rightarrow$ (low albite + quartz)

Analysis of this sequence in terms of the Ostwald rule of stages is complicated, because of the complex chemistry of clinoptilolite in natural systems. The transformation rate of the assemblage (clinoptilolite (+ mordenite) + opal-CT) to (low albite + quartz) progresses very slowly within the YM stratigraphy. The time frame is of the order of millions of years, in conformity with independent reasoning by Dibble and Tiller (1981). However, with elevation of temperature, the kinetics of transformation is drastically accelerated as cited by Barrer (1982) in an analogous synthetic zeolitic system at 175°C, where metastable intermediates persisted for only 48 hours.

3.1   *Effects Related to Emplacement of Heat-Generating Nuclear Waste*

At Yucca Mountain, the emplacement of radioactive waste will lead to transient desiccation of the host rocks surrounding the emplacement drifts due to the release of heat during radioactive decay of the waste. Eventually, the desiccated region will be rehydrated. Thermal-hydrological-chemical processes involving the desiccation, condensation, and wetting fronts in this system produces conditions involving supersaturation with respect to siliceous phases and carbonates. The distribution of precipitating mineral phases in this system is strongly related to differences in solubility as a function of temperature. Precipitation of amorphous silica is expected to be confined to a narrow zone where evaporative concentration from boiling exceeds its solubility. In contrast, calcite may precipitate in fractures over a wider zone of elevated temperature, because of its lower solubility at higher temperatures. It is anticipated that the system will respond according to the Ostwald Rule of Stages.

In an experiment set up as an analogy to THC processes in the fractured tuff at Yucca Mountain, rapid precipitation of amorphous silica precipitation was produced by passing water through crushed tuff at 80°C and then allowing it to drain into an artificial fracture where the temperature was held in a gradient from 80°C to 130°C (Dobson et al., 2003). Results of ractive transport modeling of the experiment indicated that extremely high reactive surface areas were required (approximately equivalent to that characteristic of colloidal silica) to capture precipitation rates in this system.

Precipitation of amorphous silica, calcite, and gypsum (possibly anhydrite initially) have also been observed as products of boiling in the deep underground thermal tests at Yucca Mountain (e.g., Levy et al, 1998). Modeling the distribution of mineral precipitation in the fractures in the experiments also requires extremely high surface areas to capture rapid nucleation at the boiling front. Because the rock becomes completely dry thus resulting in complete precipitation of all silica, and the rate of advance of the boiling front is captured by temperature measurements and modeling quite accurately, the rates of silica precipitation can be estimated. However, the extent to which aqueous silica becomes supersaturated with respect to amorphous silica is unknown because fluid samples extracted from rocks by ultracentrifugation have equilibrated with a lower solubility silica polymorph at lower temperatures.

4   CONCLUSIONS

In summary, the issues requiring quantification and development of a realistic treatment in reactive chemical transport models are:

- A process incorporating the Ostwald Rule of Stages, thereby allowing the model to select automatically that phase (or those phases) that should precipitate from solution under specific conditions, rather than forcing the model to precipitate an arbitrarily selected phase. This feature is particularly important when two or more phases with differing thermodynamic stabilities may be precipitating simultaneously.

- A model that incorporates nucleation, growth, and Ostwald ripening. The model should allow for varying rates of precipitation, and the progressive reduction of growth sites with time.

REFERENCES

Abercrombie, H.J., Hutcheon, I.E., Bloch, J.D. & de Caritat, P. 1994. Silica activity and the smectite-illite reaction. *Geology* 22:539-542.

Barrer, R.M. 1982. *Hydrothermal Chemistry of Zeolites*. Academic Press, New York, 360 p.

Bish, D.L. & Vaniman, D.T. 1985. Mineralogic summary of YM, Nevada. *Los Alamos National Laboratory Report LA-10543-MS*, 55 p.

Carey, J.W., Chipera, S.J., Vaniman, D.T., Bish, D.L., Viswanathan, H.S., & Carter-Krogh, K. 1997. Three-Dimensional Mineralogic Model of Yucca Mountain, Nevada, Rev. 1. Report SP344BM4. Los Alamos National Laboratory.

Carlos, C., 1989. Fracture-coating minerals in the Topopah Spring member and upper tuff of Calico Hills from Drill Hole J-13. Los Alamos National Laboratory Report LA-11504-MS, 20 p.

Carlos, C., Bish, D.L. & Chipera, S.J. 1991. Fracture-lining minerals in the lower Topopah Spring tuff at YM. *IHLWC'91*.

Carlos, C., Chipera, S.J. & Bish, D.L 1995. Distribution and chemistry of fracture-lining minerals at YM, Nevada. *Los Alamos National Laboratory Report LA-12977-MS*, 92 p.

Chang, J. & Yortsos, Y.C. 1994. Lamination during silica diagenesis-effects of clay content and Ostwald ripening. *Am. J. Sci.* 294:137-172.

Chipera, S.J. & Bish, D.L. 1988. Mineralogy of Drill Hole UE-25p#1 at YM, Nevada. *Los Alamos National Laboratory Report LA-11292-MS*, 24 p.

Dibble, W.E. Jr. & Tiller, W.A. 1981a. Non-equilibrium water/rock interactions. I. Model for interface-controlled reactions. *Geochim. Cosmochim. Acta* 45:79-92.

Dibble, W.E., Jr. & Tiller, W.A. 1981b. Kinetic model of zeolite paragenesis in tuffaceous sediments. *Clays & Clay Minerals* 29:323-330.

Dobson, P.F., T.J. Kneafsey, E.L. Sonnenthal, N.F. Spycher, and J.A. Apps 2003. Experimental and numerical simulation of dissolution and precipitation: Implications for fracture sealing at Yucca Mountain, Nevada. *Journal of Contaminant Hydrology* 62-63:459-476.

Duffy, C.J., 1993. Preliminary conceptual model for mineral evolution at YM. *Los Alamos National Laboratory Report LA-12708-MS*, 46 p.

Kastner, M. 1979. Silica Polymorphs. In R.G. Burns (ed.), *Marine Minerals, Chapter 3, Mineralogical Society of America Short Course Notes* 6:99-109. Mineralogical Society of America, Washington, DC.

Levy, S., Chipera, S., & Snow, M. 1998. Mineralogical products of the ESF Single-Heater Test. *YMP Milestone SPY147M4*. Los Alamos National Laboratory.

Morse, J.W. & Casey, W.H. 1988. Ostwald processes and mineral paragenesis in sediments. *Am. J. Sci.* 288(6):537-560.

Ostwald, W. 1897. Studien über die Bildung und Umwandlung fester Körper: *Zeitschrift für Physikalische Chemie* 22:289–330.

Parks, G.A. 1990. Surface energy and adsorption at mineral-water interfaces: An introduction. *Rev. Mineral.* 23(Mineral-Water Interface Geochem.):133-75.

Schmelzer, J. 1998. Ostwald's rule of stages. The effect of elastic strains and external pressure.: Moeller, Joerg; Gutzow, Iwan. *Z. Phys. Chem. (Munich)* 204(1/2):171-181.

Smyth, J. R. 1982. Zeolite stability constraints on radioactive waste isolation in zeolite-bearing volcanic rocks: *J. Geol.* 90:195-202.

Steefel, C.I. & Van Cappellen, P. 1990. A new kinetic approach to modeling water-rock interaction: The role of nucleation, precursors, and Ostwald ripening. *Geochim. Cosmochim. Acta* 54:2657-2677.

Vaniman, D.T. 1994. Calcite deposits in fractures at YM, Nevada. *HLRWM-94*.

Van Santen, R. A. 1984. The Ostwald step rule. *J. Phys. Chem.* 88(24):5768-5769.

Williams, L.A., Parks, G.A. & Crerar, D.A. 1985. Silica diagenesis, I. Solubility controls. *J. Sed. Petrol.* 55(3):301-311.

4.18   Reactive Transport Modeling of Biologic Iron(III) Reduction Under Varied Hydrologic Conditions

W.D. Burgos[1] and G.T. Yeh[2]

[1]*The Pennsylvania State University, Department of Civil and Environmental Engineering, University Park, PA, wdb3@psu.edu*

[2]*University of Central Florida, Department of Civil and Environmental Engineering, Orlando, FL*

Previously we demonstrated that diagonalized reaction-based models could be applied to batch systems in which parallel kinetic reactions were operative if separate experiments were used to independently formulate and parameterize kinetic rate expressions (Burgos et al., 2003). Currently our objective is to demonstrate that reaction-based models can accurately simulate complex biogeochemical systems under advective flow conditions using rate formulations/parameters obtained from independent batch experiments. For demonstration purposes we selected biological iron(III) reduction in natural sediments. All experiments were performed with a hematite-rich sand from Eatontown, NJ (366 umol DCB-extractable Fe g^{-1} sediment), and the dissimilatory metal-reducing bacterium (DMRB) *Shewanella putrefaciens* CN32 with lactate as the sole electron donor in a buffered (pH = 6.8) background electrolyte (BE) containing 45 mM PIPES, 1 mM $CaCl_2$, 0.1 mM NH_4Cl, 0.01 K_2HPO_4, 0.01 mM $MgSO_4$, and 0.1 g L^{-1} yeast extract (growth conditions). A series of batch kinetic experiments were performed to systematically measure iron reduction as a function of iron oxide surface area, cell density, and electron donor concentration in order to independently formulate a kinetic rate expression for the biological iron reduction. Batch experiments were conducted with variable sediment concentrations (0.007 – 2.0 g mL^{-1}), variable initial DMRB concentrations (10^6 – 10^9 cells mL^{-1}), and variable lactate concentrations (1 – 50 mM). Reactors were incubated at $20^{\circ}C$ on a shaker table for 7 d, and kinetic data were collected on a logarithmic scale. For each time point, soluble Fe(II) (0.2 um), 0.5 N HCl Fe(II), lactate, acetate, and pH were measured. At the final 7 d point, cell concentration (by AODC following 24 h 20 mM Na_2HPO_4 pH 7.4 extraction) and iron mineralogy (by Mössbauer spectroscopy) were also measured. For all batch experiments the overall reaction rate of Fe(II) production was zero-order (R^2 values for [0.5 N HCl Fe(II)]-vs-time ranged from 0.905 to 0.999). The overall Fe(II) production rate was directly dependent on sediment concentration (e.g., proportional to reactive iron surface area) and not dependent on lactate concentration (i.e., system was never electron donor limited). The effect of DMRB concentration on the overall Fe(II) production rate was complex and non-linear, suggesting a growth term may be required to accurately capture this effect.

Eatontown sand was carefully wet-packed into 1-cm diameter glass chromatography columns with a packed bed length of ca. 7.5 cm. Dry sand was added in 1.0 g lifts into BE containing the 10^8 cells mL^{-1} DMRB, and the overlying water height was maintained at 1.0 cm. All column experiments were conducted in triplicate or quadruplicate and run for 21 d. The only experimental variable was flow rate, which ranged from 0.4 to 10 pore volumes day^{-1}, corresponding to Darcy velocities of 7.5 to 90 cm day^{-1}. Column effluent samples were collected daily and used to measure soluble Fe(II), lactate, and acetate. At the final 21 d point, columns were deconstructed in an anaerobic chamber to collect sediment samples as a function of column length. These samples were used to measure 0.5 N HCl Fe(II), cell concentration, and iron mineralogy by Mossbauer spectroscopy. In most experiments, the effluent biogenic Fe(II) concentrations increased over the first 10 days and then remained relatively constant (Figure 1). From these pseudo steady state conditions the biogenic flux of Fe(II) (umol Fe(II) d^{-1}) was calculated as the product of the "plateau" concentration (μmol L^{-1}) times the average flow rate (L d^{-1}). These flux results clearly demonstrate that hydrologic conditions effect biologic reactions (Figure 2). Based on calculations of the Peclet number for the various flow rates, the increase in biogenic flux occurs when transport processes switch from diffusion-controlled to advective-controlled. These results highlight the intriguing linkages between hydrology and biogeochemistry.

A simple reaction network was proposed to describe these systems and included four reactions: 1 – biological reduction of hematite coupled to lactate oxidation; 2 – sorption of biogenic Fe(II) to hematite; 3 – surface "hydration" of hematite to produce and account for "free" surface sites; and, 4 – organic Pipes buffering of the solution. The batch reaction-based model BIOGEOCHEM (Fang et al., 2003) was used to simulate all batch experiments and formulate and parameterize the hematite bioreduction reaction. An empirical rate formulation that was first-order with respect to free surface sites was selected for this reaction. Fe(II) sorption was modeled using a Freundlich isotherm. Surface hydration was modeled based on measured surface area and assumed site density. Pipes buffering was modeled based on the pKa value provided by the chemical manufacturer. With no further modifications, these rate formulations/parameters and equilibrium constants were used to model the constructed column results. The reaction-based reactive transport model HYDROGEOCHEM version 5.0 (Yeh et al., 2004) was used for these simulations. These column simulations revealed that the independently obtained reaction-based parameters could only accurately capture some of the experimental results. Therefore, our current effort is to re-examine and possibly expand the reaction network to include an appropriate term for DMRB growth.

REFERENCES

Burgos, W.D., Y.L. Fang, R.A. Royer, G.T. Yeh, J.J. Stone, B.H. Jeon and B.A. Dempsey. 2003. Reaction-Based Modeling of Quinone-Mediated Bacterial Iron(III) Reduction. Geochim. Cosmochim. Acta. 67:2735-2748.

Fang, Y., G.T. Yeh, and W.D. Burgos. 2003. A general paradigm to model reaction-based biogeochemical processes in batch systems. Wat. Resour. Res. 39:1083-1108.

Yeh, G.T., J.T. Sun, P.M. Jardine, W.D. Burgos, Y.L. Fang, M.H. Li and M.D. Siegel. 2004. HYDROGEOCHEM 5.0: A Three-Dimensional Model of Coupled Fluid Flow, Thermal Transport, and HYDROGEOCHEMical Transport through Variable SaturatedConditions - Version 5.0. ORNL-****, Oak Ridge National Laboratory, Oak Ridge, TN 37831.

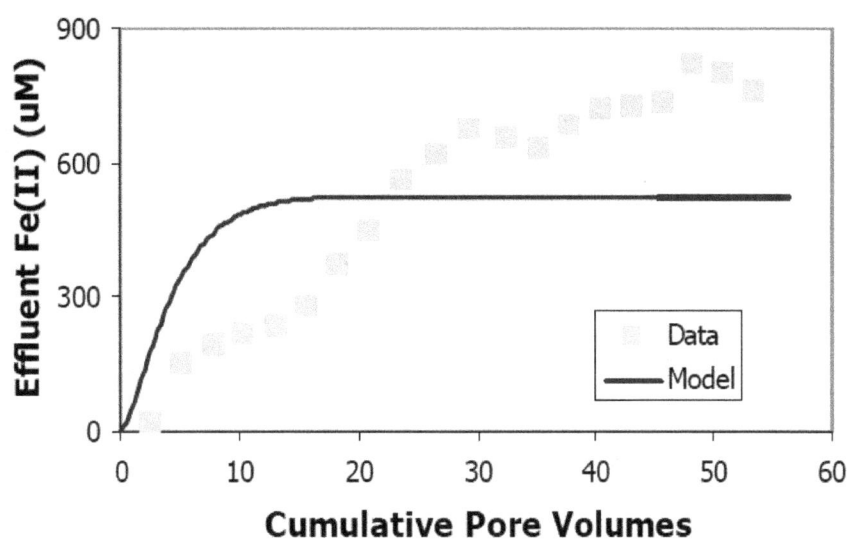

Figure 1. Effluent biogenic Fe(II) versus cumulative pore volumes from constructed columns of Eatontown hematite. Columns were 1-cm dia, 7.5-cm bed length, contained 9.0 g dry sediment, and were wet-packed with 10^8 cells/mL *Shewanella putrefaciens* CN32. Data shown is for a flow rate of 2.6 pore volumes per day.

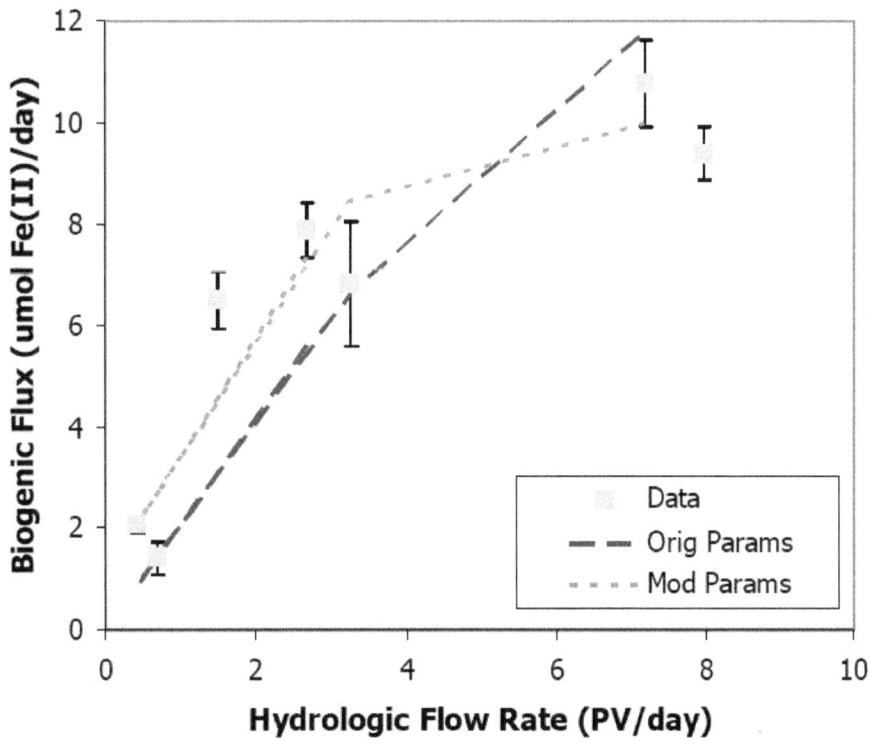

Figure 2. Psuedo-state biologic flux as a function of hydrologic flow rate.

4.19 Surface chemical and thermodynamic controls on bacterial metal reduction in subsurface environments

E.E. Roden[1]

[1]The University of Alabama, Tuscaloosa, AL, eroden@bsc.as.ua.edu.

1 INTRODUCTION

Understanding the geochemical controls on microbial reduction of soluble and solid-phase metals is a prerequisite for development of reactive transport models of inorganic contaminants and radionuclides in subsurface environments. Fe(III) oxides are ubiquitous components of soils and sediments, and often provide the largest source of electron-accepting capacity for bacterial metal reduction in subsurface environments (Lovley, 1991). Fe(III) oxides occur in a variety of phases in natural environments, ranging from poorly crystalline ferrihydrite to well-crystallized minerals such as goethite and hematite (Cornell and Schwertmann, 1996). The studies described herein were designed to explore how the surface chemical (e.g. specific surface area) and thermodynamic properties (e.g. oxide reduction potential) of Fe(III) oxides influence the rate and extent of bacterial Fe(III) oxide reduction. The results permit development of conceptual and quantitative models of enzymatic Fe(III) oxide reduction at circumneutral pH that can be compared and contrasted with models of abiotic mineral dissolution and soluble metal reduction.

2 MATERIALS AND METHODS

2.1 Fe(III) oxide phases

The Fe(III) oxides employed in this study included a variety of pure synthetic phases (Roden, 2003) as well as three previously characterized Fe(III) oxide-bearing subsoil or subsurface materials (HC, CP, and Oyster). The HC and CP materials are Fe(III) oxide/layered silicate mixtures obtained from Ultisols in Tennessee and North Carolina, respectively. The Oyster material is Fe(III) oxide-coated sand from Pleistocene Age Atlantic Coastal Plain sediments. More detailed descriptions of the properties of these materials are available elsewhere (Zachara et al., 1989; Zachara et al., 1995; Roden, 2004). The synthetic Fe(III) oxides were freeze-dried and passed through a 100-μm sieve, and their specific surface area determined by multipoint BET N_2 adsorption. The natural materials were air dried and passed through a 2 mm sieve prior to use in experiments.

2.2 Bacterial reduction experiments

The synthetic and natural Fe(III) oxide-bearing solids were suspended in sterile, anaerobic Pipes buffer (10 mM, pH 6.8) contained in sealed serum vials to obtain a Fe(III) concentration of ca. 10 mmol L^{-1}. Other experiments were conducted with soluble Fe(III) citrate (50 mM) or U(VI)-carbonate (0.01-1 mM) as electron acceptors. The electron donor for metal reduction was either 100% H_2 in the vial headspace, or 10 mM of either sodium lactate or sodium acetate. The medium was inoculated with ca. 10^8 cells mL^{-1} of either tryptic soy broth-grown *Shewanella putrefaciens* strain CN32 cells, or acetate/fumarate-grown *Geobacter sulfurreducens* cells. Samples for determination of dissolved (0.2-μm filtration and Ferrozine analysis) and total Fe(II) (0.5M HCl extraction and Ferrozine analysis) and pH were collected at 1-10 d intervals.

2.3   *Ascorbate and AH₂DS reduction experiments*

The synthetic and natural Fe(III) oxide-bearing solids were suspended in anaerobic 10 mM ascorbic acid or 10 mM AH₂DS (the reduced form of AQDS, anthroquinone-2,6-disulfonate, prepared by reacting AQDS with 100% H_2 gas in the presence of a palladium catalyst). The oxide suspensions were incubated at room temperature on a rotary shaker (250 rpm) and samples were removed with a N_2-flushed plastic syringe at regular intervals. A portion of the sample was passed through a 0.2-μm filter into Ferrozine for Fe(II) analysis, and the remainder used for determination of pH. The final (after reduction ceased) dissolved Fe(II) concentration and pH values achieved in the AH₂DS reduction experiments were used in conjunction with the E_h^0 of the AQDS/AH₂DS couple (+0.23V; Clark, 1960) to estimate the reduction potential (E_h^0) of the synthetic and natural Fe(III) oxide phases.

3   RESULTS AND DISCUSSION

3.1   *Initial rates of reduction*

Initial surface area-specific rates of bacterial reduction of the synthetic Fe(III) oxides were independent of oxide surface area and E_h^0 (Fig. 1A). In contrast, rates of abiotic Fe(III) oxide reduction by ascorbate and AH₂DS were significantly correlated with oxide surface area and E_h^0 (Fig. 1B).

3.2   *Long-term extent of reduction*

Similar patterns of Fe(II) accumulation were observed during long-term (3-week incubation) reduction of the various synthetic Fe(III) oxides by *G. sulfurreducens* (Fig. 2A). Measurements of dissolved Fe(II) and pH during the reduction experiments, together with the estimated E_h^0 values for the different Fe(III) oxides, were used to compute the free energy of Fe(III) oxide reduction for the different synthetic phases during the course of the reduction experiments. The results (Fig. 3) indicate that the cessation of oxide reduction activity could not be attributed to free energy constraints posed by dissolved Fe(II) accumulation and pH increase, because reduction rates approached zero at ΔG_{rxn} values substantially lower than the theoretical minimum of ca. -20 kJ mol^{-1} required for energy conservation during biological energy metabolism (Schink, 1997).

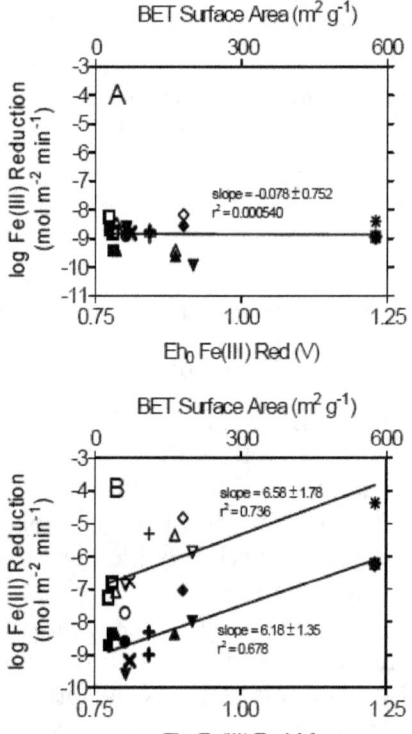

Figure 1. Initial (2-3 day incubation) surface area-specific rates of bacterial (A) and abiotic (B) reduction of synthetic Fe(III) oxides. Different symbols correspond to different Fe(III) oxide phases. Open and closed symbols in panel A refer to data for *S. putrefaciens* and *G. sulfurreducens*, respectively. Open and closed symbols in panel B refer to data for AH₂DS and ascorbic acid reduction, respectively.

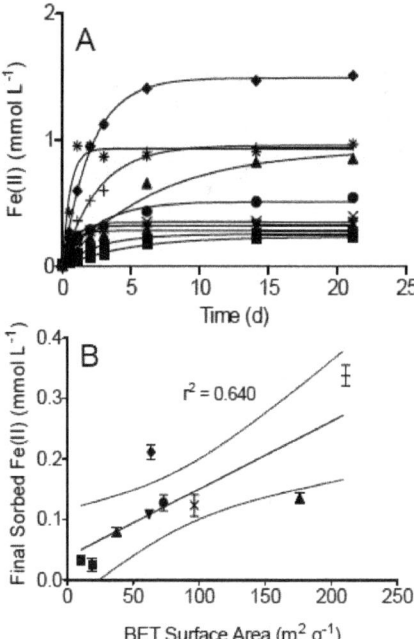

3.3 Reduction of natural Fe(III) oxide phases

Data from long-term experiments on bacterial and abiotic reduction of natural Fe(III) oxides (Fig. 4) were interpreted in relation to a standard a generalized rate law for mineral dissolution

$$J_t/m_0 = k'(m/m_0)^\gamma \qquad (1)$$

where J_t is the rate of dissolution and/or reduction at time t, m_0 is the initial mass of oxide, and m/m_0 is the unreduced or undissolved mineral fraction) in order to evaluate changes in the apparent reactivity of Fe(III) oxides during long-term biological vs. chemical reduction. The natural Fe(III) oxide assemblages demonstrated significant changes in reactivity during long-term abiotic reductive dissolution (Fig. 4A), as indicated by γ values in excess of 1 for curve-fits of the data to the generalized rate law. Much larger changes in

Figure 2. Total (0.5M HCl extraction) Fe(II) production (A) and final sorbed Fe(II) (total minus dissolved Fe(II)) concentration during reduction of synthetic Fe(III) oxides by *G. sulfurreducens*. Different symbols correspond to different Fe(III) oxide phases. Lines in panel A show nonlinear regression fits of the data to the an equation depicting the accumulation of end-product of a first-order reaction. Line in panel shows result of a linear least-squares regression analysis.

reactivity were estimated for the bacterial reduction experiments (Fig. 4B, solid symbols). However, when the analysis was restricted to the long-term "microbially-reducible" fraction of the Fe(III) oxide content of the natural solids (Fig. 4B, open symbols), the data could be well-approximated with γ values of ca. 1,

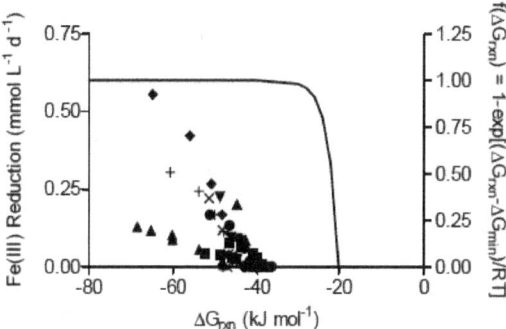

Figure 3. Rate of Fe(III) reduction during long-term *G. sulfurreducens* experiments as a function of the estimated free energy of reaction (ΔG_{rxn}). Fe(III) reduction rates were computed from the nonlinear curve-fits shown in Fig. 2. The solid line shows the relationship between reaction rate and energetics predicted based on transition state theory Lasaga (1998).

i.e. by a first-order rates process. Kinetic and thermodynamic considerations indicated that neither the abundance of electron donor (lactate) nor the accumulation of aqueous end-products of oxide reduction (Fe(II), acetate, dissolved inorganic carbon) are likely to have posed significant limitations on the long-term kinetics of oxide reduction. Rather, accumulation of biogenic Fe(II) on residual oxide surfaces appeared to play a dominant role in governing the long-term kinetics of bacterial crystalline Fe(III) oxide reduction. This assertion is supported by the correlation between oxide surface area the quantity of sorbed Fe(II) present at the end of the synthetic Fe(III) oxide reduction experiments (Fig. 2B).

3.4 General discussion and conclusions

The experimental findings support a conceptual model of bacterial Fe(III) oxide reduction kinetics that differs fundamentally from established models of abiotic Fe(III) oxide reductive dissolution in that oxide surface area exerts primary control on both the initial rate *and* the long-term extent of reduction. Numerical simulations of surface area-controlled biotic vs. abiotic Fe(III) oxide reduction indicate that this conceptual model can account for the psuedo-first order kinetics of reduction of the operationally-defined "microbially-reducible" fraction of the sediment Fe(III) oxide pool (Roden, 2004). The explicit surface area control of the initial rate and extent of oxide reduction leads to rate formulations for oxide reduction as a function of (i) electron acceptor abundance, and (ii) the density of dissimilatory metal-reducing bacteria (DMRB), that differ from those for reduction of soluble metals such as chelated Fe(III) and U(VI)-carbonate species. Rates of solid-phase Fe(III) oxide reduction are a linear function of reactive surface site density and a hyperbolic (Monod-style) function of DMRB. The latter relationship is analogous to the well-recognized hyperbolic dependence of abiotic reductive dissolution rate on ligand concentration (Hering and Stumm, 1990). In contrast, rates of soluble metal reduction are a hyperbolic function of metal concentration and a linear function of DMRB abundance.

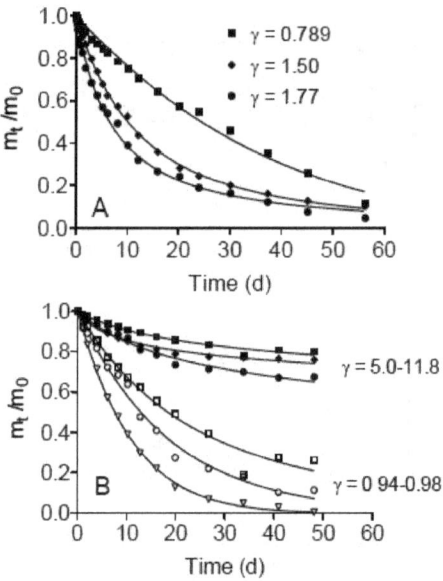

Figure 4. Kinetics of ascorbate (A) and bacterial (B) reduction of natural Fe(III) oxide-bearing materials. Squares, diamonds, and triangles represent the HC, CP, and Oyster materials, respectively. Open symbols in panel B refer to reduction of the "microbially-available" fraction of the oxide pool. The terms m_0 and m_t refer to the mass (concentration) of Fe(III) oxide present at the start of the experiment and at time t, respectively. Solid lines show nonlinear regression fits of the data to the integrated form of the generalized mineral dissolution rate law (Eqn. 1 in text). γ values refer to the results of the nonlinear curve fits.

REFERENCES

Clark, W. M. 1960. Oxidation-Reduction Potentials of Organic Systems. The Williams and Wilkins Company, Baltimore.

Cornell, R. M., and U. Schwertmann. 1996. The Iron Oxides. VCH, New York.

Hering, J. G., and W. Stumm. 1990. Oxidative and reductive dissolution of minerals. In M. F. Hochella, and A. F. White (eds.). Mineral-water interface geochemistry, pp. 427-464. Mineralogical Society of America, Washington, DC.

Lasaga, A. C. 1998. Kinetic theory in the earth sciences. Princeton University Press, Princeton.

Lovley, D. R. 1991. Dissimilatory Fe(III) and Mn(IV) reduction. Microbiol. Rev. 55:259-287.

Roden, E. E. 2003. Fe(III) oxide reactivity toward biological versus chemical reduction. Environ. Sci. Technol. 37:1319-1324.

Roden, E. E. 2004. Analysis of long-term bacterial versus chemical Fe(III) oxide reduction kinetics. Geochim. Cosmochim. Acta Submitted for publication.

Schink, B. 1997. Energetics of syntrophic cooperation in methanogenic degradation. Microbiol. Mol. Biol. Rev. 61:262-280.

Zachara, J. M., S. C. Smith, and L. S. Kuzel. 1995. Adsorption and dissociation of Co-EDTA complexes in Fe oxide containing subsurface soils. Geochim. Cosmochim. Acta 59:4825-4844.

Zachara, J. M., C. C. Ainsworth, C. E. Cowan, and C. T. Resch. 1989. Adsorption of chromate by subsurface soil horizons. Soil Sci. Soc. Am. J. 53:418-428.

4.20 Incorporating geomicrobial processes in subsurface reactive transport models

P. Van Cappellen A. Dale, C. Pallud, Y. van Lith, S. Bonneville, M. Thullner, A. Laverman and P. Regnier

Utrecht University, Department of Earth Sciences, Geochemistry, P.O. Box 80021, 3508 TA Utrecht, The Netherlands, e-mail: pvc@geo.uu.nl

The chemical structure and evolution of subsurface environments is to a large degree determined by the activity of the resident microbial populations (Chapelle, 2001). In particular, the distributions of redox-sensitive constituents are strongly affected by microbial metabolism, as electron transfer processes form the basis of biological energy generation. However, microbial reactions also modify acid-base and electrolyte properties of pore waters, thermodynamic driving forces, and the composition of particulate and colloidal matter in soils, sediments and aquifers. Comprehensive subsurface reactive transport modeling therefore requires mathematical expressions that predict the rates at which microbial populations consume and produce chemical constituents.

When representing microbial reaction processes in reactive transport models it is commonly assumed that microorganisms exhibit saturation kinetics with respect to the substrates they extract from their surroundings (e.g., Curtis, 2003). That is, with increasing concentration of a limiting substrate, the reaction rate ultimately reaches a maximum value. This behavior is captured by the so-called Michaelis-Menten (or Monod) rate expression. Michaelis-Menten kinetics, however, have mainly been verified for pure cultures utilizing soluble substrates. We present the results of a number of experimental studies, which confirm that the Michaelis-Menten formulation also holds for redox transformation processes carried out by complex, natural microbial communities. In addition, we show that the utilization of solid-phase Fe(III) substrates by pure cultures of iron reducing bacteria follows Michaelis-Menten saturation kinetics (Bonneville et al., 2004).

The accumulating empirical evidence therefore confirms that the Michaelis-Menten equation provides a generally valid description of the utilization of external substrates by microorganisms. Hence, it forms the logical starting point for the development of mathematical representations of geomicrobial processes in reactive transport models. We present an extended Michaelis-Menten model, which accounts for multiple substrates, changes in the microbial community structure, the presence of inhibitors, and the free energy yield of the microbial reaction pathways. Two applications of the extended Michaelis-Menten model are discussed.

In the first application, we analyze the order of utilization of terminal electron acceptors during the anaerobic degradation of sedimentary organic matter. The simulations show that competition between the organic carbon oxidation pathways depends in the first place on the values of the affinity constants for the electron donors and, to a lesser degree, on the growth yields of the heterotrophic microbial populations. However, they further indicate that, when labile organic matter is abundant, the affinity constants for the electron acceptors also play an important role in the reaction dynamics. In particular, these affinity constants influence the relative distribution of TEA utilization over competing reduction pathways. For example, a decrease in the affinity constant of Fe(III) (hydr)oxides for iron reducing bacteria may shift

Fe(III) reduction to reaction with sulfide, which consequently promotes organic matter degradation by sulfate reducing bacteria. Furthermore, with increasing availability of labile organic matter, there is growing spatial and temporal overlap among respiratory pathways using different TEAs. This theoretical result agrees with recent experimental studies (e.g., Koretsky et al., 2003).

In the second application of the extended Michaelis-Menten model, we consider the anaerobic oxidation of methane by sulfate (AOM), a geomicrobial reaction system operating close to its thermodynamic limit (e.g., Hoehler et al., 1994). The consortia of methane oxidizing archaea and sulfate reducing bacteria carrying out AOM depend on a close syntrophic association and energy sharing for their survival. In particular, the sulfate reducers are essential for the disposal of H_2 produced by the methane oxidizers. To simulate the activity of these microbes, as well as the turnover of methane at the seafloor, we have developed a reaction network model including methane oxidation and sulfate reduction coupled to H_2 oxidation, as well as methanogenesis, acetogenesis, fermentation, and sulfate reduction coupled to oxidation of organic fermentation products.

Our current reaction network model embodies the still fragmentary conceptual understanding of how AOM communities function. Yet the model already requires over 40 independent parameters. This large number of parameters reflects the inherently complex nature of reaction systems in which multiple microbial groups participate. To identify those parameters, or combination of parameters, that most significantly affect the model output, we use a two-level factorial analysis. A baseline, or nominal, simulation is first conducted, using the best available estimates of the parameters. Subsequently, perturbations are carried on groups of two or more parameters and the deviation from the nominal run of some defined model response is monitored (Montgomery, 1997). The sensitivity range for each parameter is chosen based on experimental data or informed guesswork.

Using a nominal coastal sediment scenario, the most sensitive parameters for AOM are found to be those belonging to sulfate reduction coupled to H_2 oxidation and methane production from acetate. In particular, the model response is sensitive to the Gibbs energies of reaction of the two pathways. The latter observation reinforces the importance of including thermodynamic constraints in the kinetic descriptions of the metabolic pathways in the reaction network. The sensitivity analysis further shows that the overall rate of AOM is not sensitive to the kinetic and thermodynamic parameters directly related to the methane oxidation pathway or the biomass of methane oxidizers. That is, the explicit representation of these parameters, and the related variables, does not improve the performance of the reaction network model. This result, which at first may seem counterintuitive, would not have been obtained without considering the entire set of microbial pathways and their interactions.

As knowledge about microbial processes and their interactions expands, it becomes possible to develop increasingly detailed model representations of geomicrobial reaction systems. This, in turn, creates the need for robust approaches for parameter identification and calibration, as illustrated here for the case of AOM. Global sensitivity analyses, for example using factorial analysis, help identify those parameters that need to be targeted in future laboratory and field experiments. They further provide guidelines for the derivation and parameterization of simplified mathematical reaction models that can be incorporated in reactive transport codes. Therefore, the development of detailed biogeochemical reaction network models helps closing the gap between fundamental research in geochemistry, microbial ecology and molecular biology, and application-oriented reactive transport modeling.

REFERENCES

Bonneville, S., Van Cappellen P. & Behrends, T. 2004. Microbial reduction of iron(III) oxyhydroxides: Effects of mineral solubility and availability. *Chemical Geology* (submitted).

Chapelle, F.H. 2001. *Ground-Water Microbiology and Geochemistry*. New York: John Wiley & Sons.

Curtis, G.P. 2003. Comparison of approaches for simulating reactive solute transport involving organic degradation reactions by multiple terminal electron acceptors. *Computers & Geosciences* 29: 319-329.

Hoehler, T.M., Alperin, M.J., Albert, D.B., & Martens, C.S. 1994. Field and laboratory studies of methane oxidation in an anoxic marine sediment - evidence for a methanogen-sulfate reducer consortium. *Global Biogeochemical Cycles* 8: 451-463.

Koretsky, C.M., Moore, C.M., Lowe, K.L., Meile, C., DiChristina, T.J. & Van Cappellen, P. 2003. Seasonal oscillation of microbial iron and sulfate reduction in saltmarsh sediments (Sapelo Island, GA, USA). *Biogeochemistry* 64: 179-203.

Montgomery, D.C. 1997. *Design and Analysis of Experiments, 4th Ed*. New York: Wiley.

4.21 Testing Coupled Microbial and Reactive Solute Transport Models with In Situ Experiments: REX and Redox Zone Experiments at Äspö (Sweden)

J. Samper[1], J. Molinero[1], Ch. B. Yang[1] and G. Zhang[2]

[1]*Escuela de Caminos, Universidad de Coruña, Coruña, Spain, jsc@iccp.udc.es*

[2]*Earth Sciences Division. Lawrence Berkeley Nat. Lab. Berkeley, CA-94720*

1 INTRODUCTION

Underground facilities are being operated by several countries around the world for performing research and providing demonstration of the safety of deep radioactive waste repositories. The Äspö Hard Rock Laboratory is one of such facilities launched and operated by the Swedish Nuclear Waste Management Company where various in situ experiments have been performed in fratured granites. The Äspö HRL is located in the southeast part of Sweden, 400 km south of Stockholm. The underground facility consists of a 3,600 m long tunnel which starts with an access ramp and runs into two turns down to a depth of 450 m under the Äspö island (Molinero and Samper 2003). Here we present coupled reactive and microbial transport models for the REX and Redox Zone Experiments. The REX experiment was performed at a packed section of a borehole drilled from the access tunnel and aimed at studying and evaluating the mechanisms of oxygen depletion near the access tunnel. The Redox Zone Experiment aimed at evaluating the large-scale effects of the construction of the access tunnel on the hydrochemical conditions of a fracture zone.

2 MICROBIAL PROCESSES AT THE ÄSPÖ SITE AND THE REX EXPERIMENT

Safety assessment of the Swedish concept for high-level radioactive waste disposal considers the stability of redox conditions in the engineered barriers (copper, bentonite and backfill) and the surrounding rock as a key factor for the long-term performance of the repository. If a cooper canister corrodes, radionuclides may be released. Biochemical reactions are catalyzed by microorganisms, so it is needed to define these biochemical processes in relation to electron-donor availability and microbial diversity in deep groundwaters.

The Microbe-REX Project of the Äspö site (Kotelnikova & Pedersen, 1999) investigated microbial nutrients, organism diversity, microbial activity and O_2 reduction potential. This project studied O_2 depletion processes by creating a controlled oxidizing perturbation in a deep rock environment at the Äspö site. Oxygen was injected into a fracture zone and water samples were collected for microbiological analyses. The REX niche is located 380 m below ground surface. The experiment was performed in a closed re-circulating system placed in borehole KA2861A. The circulation loop had a total volume of about 1 L, and before each O_2 pulse it was filled with fresh O_2-free groundwater from the adjacent borehole KA2862A. O_2 injection pulses started by replacing part of the volume by groundwater samples that had been previously saturated with either air or $O_2(g)$ (Figure 1).

The results of this experiment indicate that groundwaters contain abundant and diverse microbial populations, including not only anaerobic microorganisms, but also facultative aerobic and microaerophilic organisms. Several microbially-catalyzed reactions may contribute to biomass production in accordance to available nutrients for biomass production.

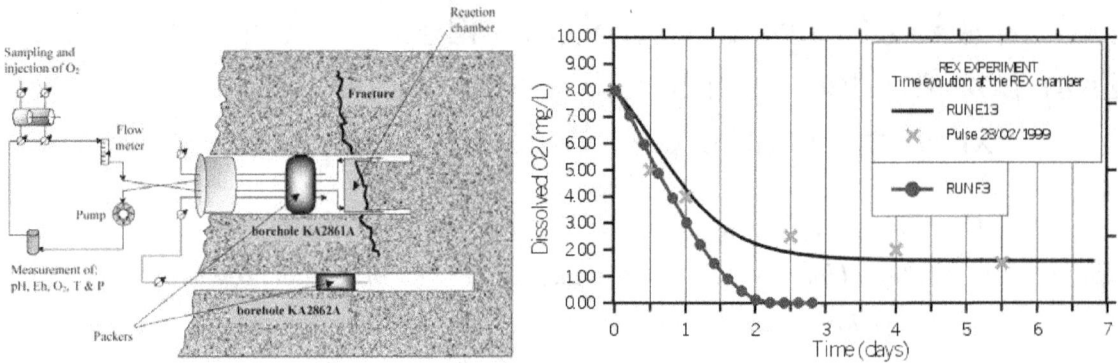

Figure 1. Setup of the REX in situ experiment (left) and time evolution of measured and computed dissolved oxygen (right). Computed values are shown for models considerin: DOC respiration (Run E13) and the combined effect of DOC respiration and methane oxidation (Run F3).

Both attached and dissolved microorganisms may be responsible for the observed oxygen depletion under in situ conditions. Although oxygen injection may inhibit iron- and sulfate-reducing bacteria, these microorganisms still appear with pronounced numbers. The percentage of iron-reducing organisms increases from 1.2-2.3% in the borehole to 0.2-55% in the REX chamber after a low oxygen pulse. The viable counts of sulfate- and iron-reducing bacteria are negatively correlated with increasing redox potential (Kotelnikova & Pedersen, 1999).

A wide variety of microorganisms, including fungi, facultatively anaerobic bacteria, and strict anaerobes are capable of reducing iron. Some of these organisms are capable of oxidizing organic compounds to CO_2, while others are incomplete oxidizers. Iron-reducing bacteria are known to catalyze the oxidation of aromatic compounds. Microbiological investigations in the Microbe-REX Project indicate that fermentation is a common microbial process taking place in the fracture zone which is carried out by anaerobic *rods* and *cocci*. Deep in the fracture zone, HS^- is unstable and is oxidized by Fe(III) or Mn(IV).

A coupled biogeochemical reactive transport model was performed for the REX experiment. The model reproduces the trends of measured oxygen consumption (Figure 1). Aerobic respiration of organic matter seems to be the main responsible process of oxygen uptake. The model indicates that abiotic consumption of dissolved oxygen in air-saturated water would take place on about 500 years. However, this time is reduced to 0.5 year when microoibally-driven DOC respiration and methane oxidation are considered in the model.

3 NUMERICAL MODEL OF THE REDOX ZONE EXPERIMENT

On March 13th, 1991 the access tunnel of the Äspö HRL intersected a vertical fracture zone at a depth of 70 m below sea level. This vertical fracture zone is known within the context of the Äspö HRL as the Redox Fracture Zone because it was used to perform a long-term experiment for evaluating the effects of the construction of the access tunnel on the hydrogeological, hydrochemical and redox conditions prevailing at this fracture zone (Banwart et al., 1996). The fracture zone is approximately vertical with a thickness of 1 m. Prevailing hydraulic conditions changed drastically first when the access tunnel intersected the fracture zone and later when borehole KR0013B was open to flow. At the vicinity of the tunnel, dilution of the initially saline groundwaters by fresh recharge water was the dominant process controlling the hydrochemistry evolution during the experiment. Most chemical species showed dilution trends except for dissolved bicarbonate and sulfate which increased with time (Molinero & Samper 2003).

Molinero & Samper (2003) presented a hydrogeological groundwater flow and solute transport model of the experiment which is consistent with head and salinity data collected prior to tunnel construction and reproduced simultaneously the observed drawdowns and dilution curves induced by the construction of the tunnel. The revised hydrogeological model provided the approriate framework to perform simulations of coupled groundwater flow, reactive solute transport and microbial processes of the Redox Zone Experiment.

The reactive transport numerical model accounts for more than 60 homogeneous reactions, including aqueous complexation, acid-base, gas dissolution and redox processes, as well as 5 heterogeneous reactions including mineral dissolution / precipitation and cation exchange. A reactive solute transport model without microbial processes reproduces observed concentrations of most dissolved species both before and after tunnel construction, but fails to match measured bicarbonate and sulfate data.

The microbial model accounts for the fermentation of particulate organic carbon (POC) and the oxidation of dissolved organic carbon (DOC). POC containing sulphur compounds ferments in the shallow anaerobic zone releasing DOC and sulphur compounds. POC has a low mobility due to its large molecular weight. The low mobility of POC in the model is achieved by assigning a large distribution coefficient of 100 L/g to this species. Then, DOC is transported to deeper parts of the fracture zone and is further degraded by heterotropic bacteria which employ Fe^{3+} as electron acceptor. The model uses yeast as the microbial species representing all the fermentation-workers. POC acts as a substrate for yeast. A single-Monod kinetics is employed to describe POC fermentation. DOC and HS^- are products of POC fermentation. The microbial model assumes that SO_4^{-2} is a direct product of fermentation. As a product of fermentation, DOC is further transported into the deep parts of the fracture zone and degraded by heterotrophic bacteria such as iron-reducing bacteria. Fe^{3+} is considered to be the only electron acceptor to account for the biodegradation of DOC. Taking DOC as a substrate and Fe (III) minerals as electron acceptors, iron-reducing bacteria grow yielding HCO_3. Fe (III) minerals are abundant in the granitic fracture zones of the Äspö site, so it is assumed that the growth of the iron reducer is not limited by Fe^{3+}.

The numerical model is solved with BIO-CORE (Zhang & Samper, 2001), a general-purpose finite element solver developed at the University of A Coruña. BIO-CORE copes with both thermodinamically-controlled abiotic geochemical reactions and subsurface microbial processes in 2-D nonisothermal partly or fully saturated porous media and considers the availability of substrates for attached microorganisms by coupling a diffusion layer model to account for biofilm resistance.

Specific growth rates, half-saturation constants, yield and proportionality coefficients, and POC and DOC initial and boundary concentrations were estimated by trial-and-error matching of breakthrough curves of bicarbonate, sulfate, DOC and pH at a depth of 70 m. Computed HCO_3^- and SO_4^{2-} concentrations are extremely sensitive to small changes in microbial parameters. Calibration of microbial parameters required a lot of effort and presented numerous convergence problems. Calibrated microbial parameters are reported by Molinero et al., 2004).

Figure 2 shows the comparison of measured and computed concentrations of bicarbonate and sulfate near the tunnel location at a depth of 70 m. This figure also includes the values obtained with a reactive transport model that ignores microbial processes. Such model predicts a mild increase in bicarbonate and a decrease in sulfate. It clearly fails to reproduce the observed trend of increasing concentrations. On the other hand, the model accounting for microbial processes reproduces the observed increase in bicarbonate and sulfate during the experiment. In adition, values of DOC computed with this model are consistent with measured DOC concentrations. Computed pH values remain near neutral which are also consistent with field observations of pH remaining stable. Model sensitivity was evaluated for specific growth rates, half-saturation constants, proportionality and yield coefficients as well as initial and boundary POC and DOC concentrations (Samper et al., 2004). Results of the sensitivity analysis indicate that computed concentrations of bicarbonate and sulfate are sensitive to changes in the initial concentration of POC and the boundary concentration of DOC, but they lack sensitivity to the initial concentration of DOC and the boundary concentration of POC (Molinero et al., 2004).

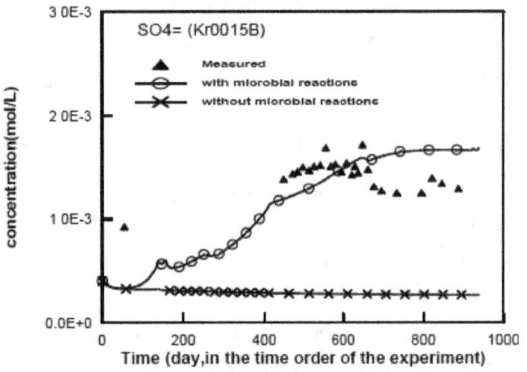 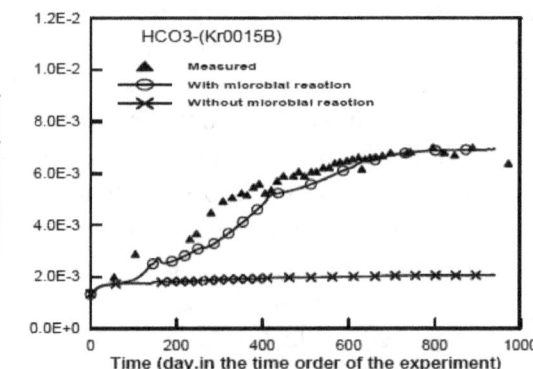

Figure 2. Comparison of measured and computed evolution of bicarbonates (left) and sulfates (right) with a model accounting for microbial processes (solid lines) and a model ignoring these processes (dashed lines).

4 CONCLUSIONS

Coupled hydrobiogeochemical models of REX and Redox Zone Experiments have been presented. The model of the Redox Zone experimen accounts for water flow, reactive solute transport, fermentation of POC by yeast and oxidation of DOC by iron-reducing bacteria. This model provides a plausible quantitatively-based explanation for the unexpected trends of bicarbonate and sulfate at the Redox Zone Experiment. Coupled hydrobiogeochemical modeling proves to be a powerful tool to improve our understanding of the hydrochemical evolution of a geological formation affected by the construction of an underground facility similar to that planned for a future deep geological repository.

REFERENCES

Banwart, S.; Tullborg, E.-L.; Pedersen, K.; Gustafsson, E.; Laaksoharju, M.; Nilsson, A.-C.; Wallin, B. and Wikberg, P. (1996): Organic carbon oxidation induced by large-scale shallow water intrusion into a vertical fracture zone at the Äspö Hard Rock Laboratory (Sweden). Journal of Contaminant Hydrology, 21, 115-125.

Kotelnikoba, S. & Pedersen, K. (1999): The microbe REX Project: Microbial O_2 comsumption in the Äspö tunnel. SKB Technical Report. Stockholm.

Molinero, J. & Samper, J. (2003): Modeling groundwater flow and solute transport in fracture zones: conceptual and numerical models of the redox zone experiment at Äspö (Sweden). Journal of Hydraulic Research. Special Issue for Groundwater. In press.

Molinero, J.; J. Samper, C.B. Yang & G. Zhang (2004): Biogeochemical reactive transport model of the Redox zone experiment of the Äspö hard rock laboratory (Sweden), Nuclear Tecnology (accepted, in press).

Samper, J., J. Molinero, C.B. Yang and G. Zhang, (2004) "Redox Zone II. Coupled Modeling of Groundwater Flow, Solute Transport, Chemical reactions and Microbial Proceses in the Äspö island", SKB Technical Report, Stockholm, In press.

Zhang, G. & Samper, J. (2001): BIO-CORE[2D©]. A code for nonisothermal hydrobiogeochemical reactive transport. Users Manual V 0. Technical Report. University of A Coruña, Spain. 244 pp.

Modeling Thermal-Hydrologic-Chemical (THC) Coupled Processes with Application to Underground Nuclear Tests at the Nevada Test Site: A "Grand Challenge" Supercomputing Problem

P.C. Lichtner and A.V. Wolfsberg
Los Alamos National Laboratory, Los Alamos, New Mexico
lichtner@lanl.gov

1 INTRODUCTION

At the Nevada Test Site (NTS), the continental nuclear weapons testing site in the United States, 828 underground tests were conducted between 1951 and 1992 (DOE, 2000). This study is focused on the BENHAM underground nuclear test conducted on December 19, 1968. During the period from 1996 to 1998, a number of radionuclides consisting of ^{3}H, ^{137}Cs, ^{60}Co, ^{152,154,155}Eu, ^{237}Np, ^{239,240}Pu, ^{241}Am, ^{14}C, ^{36}Cl, ^{99}Tc, ^{129}I, ^{90}Sr, and ^{214}Pb were detected in trace quantities 1.3 km from the BENHAM test. The radionuclides were discovered in well ER-20-5 #3, sampling a lava formation located near the depth of the working point (WP), and well ER-20-5 #1, sampling the TSA welded tuff aquifer located some 500 m above the WP situated at a depth of 1402 m (Wolfsberg et al., 2002; Kersting et al., 1999). Kersting et al. (1999) report that isotopic fingerprinting indicated the ^{239,240}Pu originated at BENHAM, rather than the closer TYBO test executed on May 14, 1975, at a depth of 765 m, and that in both aquifers ^{239,240}Pu was associated with colloids. These observations indicate an approximate travel time for ^{239,240}Pu on the order of 30 years or shorter. The BENHAM test, with an announced official yield of 1.15 megatons (DOE, 2000), produced a spherical cavity estimated at 200 m in diameter. The bottom of the cavity filled with melted rock, referred to as melt glass. A cylindrical, rubblized, chimney formed as rock above the cavity collapsed, extending from the working point of the test to above the TSA welded tuff aquifer. The melt glass is thought to contain a large proportion of the radionuclide inventory for Pu, whereas other less refractory radionuclides are distributed between the melt glass and the surrounding chimney material (Pawloski et al., 2001). A puzzling question arises as to how ^{239,240}Pu (and other radionuclides) could have migrated from the BENHAM test to two distant wells located in aquifers that apparently do not interact with one another, except through the chimney system created by the test. In addition, the radionuclides in question have wide ranging chemical properties including sorptive, solubility, and complexing affinity. This contribution attempts to elucidate the mechanisms responsible for transporting this diverse group of radionuclides over the relatively short times and long distances observed at the NTS.

Because the geometry of the system is intrinsically 3-D, consisting of a vertical cylindrical chimney with melt glass at the bottom embedded in a horizontally layered medium with an ambient lateral flow gradient, simulations to be at all realistic must be carried out in three spatial dimensions. In addition, field length scales for the system range from 100 m corresponding to the chimney radius to 1.3 km equal to the distance from the BENHAM test WP to the observation wells. Within the cavity-chimney-melt-glass region it is necessary to resolve length scales on the order of 10 m or less to capture thermal effects resulting in highly tortuous convection cells. These considerations necessitate the use of high performance computing for all but the coarsest grids.

To explain the observed radionuclide breakthrough behavior, a complex thermal, hydrologic, and chemical (THC) coupled process model is called for. However, before embarking on this endeavor with its extensive data requirements, it is prudent to take a more simplified approach to first explore the complex interplay between heat, fluid flow, and solute transport resulting from an underground test. To this end, the focus here is on coupling heat and fluid flow processes while using a relatively simple formulation of non-reactive solute transport and calibrating the predicted breakthrough times to well data. Reserved for future studies is the additional coupling of more complex chemical processes. Nevertheless, interpretation of results obtained from the simplified model provides unique insights and raises interesting questions into system responses to a transient heat source and migration of reactive and non-reactive solutes in fractured rock aquifers.

2 RADIONUCLIDE MIGRATION FROM AN UNDERGROUND NUCLEAR TEST AT THE NEVADA TEST SITE

2.1 Conceptual Model

The conceptual model used to describe radionuclide migration from an underground nuclear test consists of a near-field region composed of the chimney-cavity-melt-glass system, and a far-field region consisting of the surrounding host rock (Pawloski et al., 2001; Wolfsberg et al., 2002). Following the detonation and rewetting of the near-field region, vigorous convection takes place in the high permeability chimney as buoyant flow driven by heat released from the melt glass moves up the chimney carrying radionuclides to the overlying aquifers. The host rock is modeled as a layered medium with effective homogeneous properties for each stratigraphic unit. The rubbleized chimney is modeled as a porous medium with relatively high permeability and high porosity. In a highly simplified description of this system, three con-

THC Coupled Processes:

$$\frac{\partial}{\partial t}\phi\rho + \boldsymbol{\nabla} \cdot \boldsymbol{F}_\rho = 0 \tag{1a}$$

$$\frac{\partial}{\partial t}\left[\phi\rho U + (1-\phi)\rho_r c_r T\right] + \boldsymbol{\nabla} \cdot \boldsymbol{F}_e = 0 \tag{1b}$$

$$\frac{\partial}{\partial t}\phi RC + \boldsymbol{\nabla} \cdot \boldsymbol{F}_C + \lambda C = 0 \tag{1c}$$

Flux: $\boldsymbol{F}_\rho = \boldsymbol{q}\rho$ \qquad (2a)

$$\boldsymbol{F}_e = \boldsymbol{q}\rho H - \kappa\boldsymbol{\nabla}T \tag{2b}$$

$$\boldsymbol{F}_C = f_c \boldsymbol{q}C - \phi D\boldsymbol{\nabla}C \tag{2c}$$

Darcy Velocity: $\boldsymbol{q} = -\dfrac{k}{\mu}\left(\boldsymbol{\nabla}p - \rho g\hat{\boldsymbol{z}}\right)$ \qquad (2d)

Symbols: temperature T, pressure p, solute concentration C, porosity ϕ, density ρ, energy U, enthalpy H, permeability k, viscosity μ, acceleration of gravity g, rock heat capacity c, thermal conductivity κ, rock density ρ_r, retardation R, collide enhancement factor f_c, dispersion/diffusion coefficient D, decay constant λ.

servation equations are needed to describe these processes: mass, energy, and solute corresponding to Eqns. (1a), (1b), and (1c). A single continuum description is employed based on effective properties for porosity, permeability, and other material quantities to represent flow and transport through a fractured porous medium [see c.f. Lichtner (2000)]. In a fractured porous medium, in the limit when full equilibration occurs between fracture and matrix continua due to small fracture spacing or slow flow rate, the equivalent continuum representation of the fractured medium is approached. For an upscaled, equivalent continuum formulation, effective porosity and permeability are related to intrinsic fracture, matrix properties (subscripts f, m), by the expressions $\phi = \epsilon_f\phi_f + (1-\epsilon_f)\phi_m$ and $k = \epsilon_f k_f + (1-\epsilon_f)k_m$, where ϵ_f refers to the fracture volume fraction. In this case, fracture and matrix porewater concentrations are equal. However, for the system under consideration it is not expected that these conditions apply due to the high flow rate and large fracture spacing. This hypothesis is tested and discussed below. Although in general, radioactive decay, retardation, and colloid velocity enhancement factors are needed to describe solute transport, in what follows a single non-reactive tracer is considered ($R = 1$). Transport of sorbing radionuclides is presumed to occur via colloids. For colloids to be effective, it would also seem necessary that sorption on colloids be irreversible, thereby minimizing competition with host rock minerals. In the presence of reversible sorption, colloids must compete with stationary minerals requiring a large colloid concentration to achieve a comparable cation exchange capacity or surface site density (Honeyman and Ranville, 2002). Presumably two pools of the sorbing radionuclides ^{137}Cs and ^{239,240}Pu are present, a highly mobile fraction that becomes irreversibly attached to colloids, and a immobile fraction that remains held to mineral surfaces within the chimney and possibly along the flow path outside the chimney. Because we are primarily interested in describing breakthrough at the ER-20-5 wells, we need only be concerned with the mobile radionuclide fraction. This fraction is presumed to consist of nonsorbable species (such as ^{3}H and ^{36}Cl), in addition to sorbable species (such as ^{137}Cs and ^{239,240}Pu). These latter species are assumed to be irreversibly sorbed onto colloids and to undergo colloid-facilitated transport with little or no retardation. Thus in either case, it can be assumed that radionuclide migration is described by the non-reactive solute transport equation.

2.2 Model Implementation

The computational domain used in the simulations measures 1.5 km × 500 m × 1050 m, encompassing both the BEN-HAM test and the ER-20-5 observation wells (Figure 1). The cavity radius (100 m), chimney height (1090 m), and melt-glass volume (3.22×10^5 m^3) are related empirically to the test yield (Pawloski, 1999). The domain is divided into a moderately sized grid with 95 × 50 × 65 nodes, with grid spacing ranging from 7 m near the melt glass to 20 m away from the source. Model parameters, considered as effective properties, used in the simulations are listed in Table 1. Zero flow boundary conditions are imposed along the four sides parallel to the ambient flow direction. Pressure boundary conditions are imposed at the up and down gradient boundaries taking into account the hydrostatic increase in

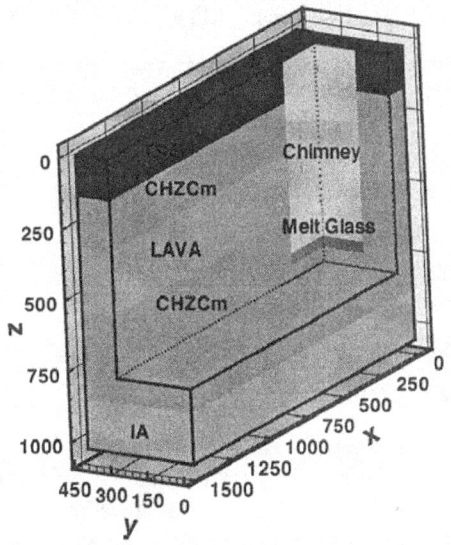

Figure 1: Cutaway view showing the grid used in the 3-D simulations. The position of the melt glass, chimney, and placement of TSA, upper CHZCm, LAVA, lower CHZCm, and IA stratigraphic units are indicated in the figure. Observation wells ER-20-5 #1 and #3, sampling in the TSA and LAVA aquifers, respectively, lie in the left y-z plane of the figure, 1.3 km from the chimney center.

pressure with depth with a geothermal gradient of 0.01°C/m. The initial pressure and temperature distribution is obtained by first running to a steady state without the chimney-cavity-melt-glass present. The initial temperature of the melt glass is set to 290°C, just below the boiling point of water at depth, mainly to avoid two-phase conditions which would greatly complicate the calculations. However, this is a reasonable starting point for this investigation because only after single-phase conditions are reached is flow away from, rather than towards, the chimney likely to occur. Investigation of the rewetting period is left for future work.

Model simulations are carried out with the initial tracer concentration normalized to the initial concentration in the melt-glass pore volume and zero outside. Calibration is achieved by matching calculated breakthough curves to field data in wells ER-20-5 #1 and ER-20-5 #3. The lateral pressure gradient and effective LAVA porosity were used as fit parameters. A scale factor for each species is applied to the field data to obtain a match to the decay corrected breakthrough curves from wells ER-20-5 #1 and #3. The value of the scale factor provides an estimate of the corresponding initial radionuclide inventory for the mobile fraction of radionuclides measured in the observation wells. Four of the twelve radionuclide species detected in the ER-20-5 observation wells consisting of ^{3}H, ^{137}Cs, ^{239,240}Pu, and ^{36}Cl were modeled. These species represent short- and long-lived radionuclides, and those transported by colloids. A (possibly nonunique) match to the field data is achieved with a lateral flow gradient of 1 m/km and an effective LAVA porosity of 0.012. Although the four radionuclides considered to date represent the range in decay rates and transport properties of the entire set of radionuclides detected in the observation wells, complete analysis will require the consideration of all radionuclides detected.

Calculations are performed using the parallel computer code PFLOTRAN, a reactive flow and transport code developed at LANL. The code PFLOTRAN describes Darcy flow coupled to heat and multicomponent mass conservation equations. Although PFLOTRAN can account multicomponent chemical reactions including mineral precipitation and dissolution, sorption, and complexing reactions, in this study a single non-reactive tracer species is considered. PFLOTRAN is based on the PETSc parallel libraries developed at

Table 1: Stratigraphy and associated physical properties for different stratigraphic units representing the host rock at the BENHAM underground nuclear test (Wolfsberg et al., 2002), with the exception of the LAVA effective porosity which is treated as a fit parameter.

Unit Name	Depth [m]	ρ_r [kg/m^3]	c_r [J/(kg K)]	κ [J/(m s K)]	ϕ —	k [m^2]
TSA	0–145	2500	880	1.8	0.005	5.00×10^{-11}
CHZCm	145–320	2350	1154	1.2	0.250	2.00×10^{-14}
LAVA	320–550	2270	1000	2.5	0.012	1.00×10^{-12}
CHZCm	550–850	2350	1154	1.2	0.250	2.00×10^{-14}
IA	850–1050	2270	1000	2.5	0.083	5.00×10^{-13}
Chimney	0–714	2354	1043	1.8	0.185	1.00×10^{-11}
Glass	679–714	2500	1154	2.7	0.010	1.00×10^{-14}

Argonne National Laboratory (Balay et al., 1997) and makes use of object-oriented features in FORTRAN 90. PETSc (latest version 2.2.0) has achieved a high level of maturity that allows rapid development of efficient parallel implementation for solving systems of non-linear partial differential equations. PFLOTRAN is essentially platform independent and can run on any machine that PETSc runs on. This includes laptop computers, workstations, and massively parallel high performance computing facilities. PFLOTRAN uses a domain decomposition approach based on the fully implicit, non-linear, iterative solver (SNES) libraries in PETSc and uses the distributed array (DA) parallel data structure to manage communication between processors. This includes the local data owned by each processor and bordering grid cells owned by neighboring processors.

The simulations involve three degrees of freedom—pressure, temperature, and solute concentration—per grid block, resulting in a total of $308,750 \times 3 = 926,250$ degrees of freedom. Some savings in computation time could be obtained by solving the tracer transport equation sequentially from temperature and pressure since the latter equations do not involve the solute concentration, however, this savings is minimal when employing parallel computation with many processors. The calculations reported here were carried out on LANL's QSC and PNNL's MPP2 machines using 64 and 128 processors. Computation times scaled almost ideally with the number of processors. The memory requirements for this moderately sized problem are minimal compared to the available memory. On MPP2 each processor has 3-4 GB of memory, compared to 4 GB per processor on QSC. The advantages of parallel computing become even more pronounced as the complexity of the problem is increased, for example, by taking into account multicomponent processes and finer grid resolution. This would allow coupling competitive reactive processes for solute sorption to matrix materials and colloids as well as colloid filtration processes as described by Wolfsberg et al. (2002) into a three-dimensional simulation framework. For example, reducing the grid spacing by a factor of two in each direction would increase the number of degrees of freedom by eight. Increasing the number of chemical components to ten results in a four-fold increase in the degrees of freedom. Finally, including dual continuum capabilities with, for example, ten matrix nodes for each fracture node leads to an overall increase of $8 \times 4 \times 10 = 320$, or approximately 300 million degrees of freedom, which is beginning to stretch the limits of available computing facilities.

2.3 Results

Shown in Figure 2 is the calculated temperature distribution in the neighborhood of the test after an elapsed time of 10 years. Thermal plumes can be seen rising from the melt glass as it cools. As can be seen in the figure, buoyancy driven flow has reached the upper TSA aquifer and has carried the thermal plume along with radionuclides laterally. The melt glass temperature becomes highly nonuniform due to convective cooling. Apparent from this figure is that early breakthrough occurs in the upper TSA aquifer, and not the

LAVA where almost no temperature increase is visible after 10 years even though the LAVA unit is much closer to the source (see Figure 1). This is a consequence of the lower permeability and higher porosity of the LAVA compared to the TSA.

Radionuclide breakthrough curves for the 3-D simulation are shown in Figure 3 for ^{3}H, ^{137}Cs, ^{36}Cl, and ^{239,240}Pu. A good match to field observations in wells ER-20-5 #1 and #3, indicated by the symbols, is obtained for ^{3}H, ^{137}Cs, and ^{36}Cl, by fitting to the tracer breakthrough curve corrected for radioactive decay. Radioactive decay is included for ^{3}H and ^{137}Cs by transforming the breakthrough concentrations by the decay factor $e^{-\lambda t}$, with λ corresponding to tritium and cesium half lives of 12.32 y and 30.2 y, respectively (radioactive decay of ^{36}Cl and ^{239,240}Pu with half lives of 301,000 years and 24,100/6560 years, respectively, is negligible over the time scale of these simulations). In particular, the coupled model is able to explain the relative magnitude in concentrations observed in the upper TSA and lower LAVA aquifers. The results of the calculations are consistent with the interpretation that species ^{3}H and ^{36}Cl move unretarded, as well as some portion of ^{137}Cs and ^{239,240}Pu sorbed irreversibly to colloids. However, they do not require that there be no retardation. For example, if a smaller value is used for the chimney porosity, then retardation (or refitting the lateral flow gradient) would be necessary to obtain the observed breakthrough times. For a flow dominated system, the porosity times the retardation factor appears in the transport equation and the two quantities cannot be determined separately [see Eqn.(1c)]. Note that this analysis does not provide an estimate of what fraction of the reactive species sorb to colloids, or what fraction of colloids filter out of the flow system.

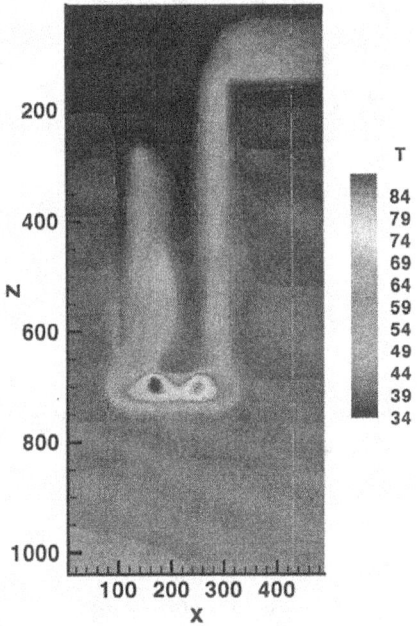

Figure 2: Near-field temperature distribution after an elapsed time of 10 years.

Transport of the species ^{239,240}Pu appears to be an anomaly. Comparison of the measured ^{239,240}Pu concentrations with the tracer breakthrough curves would suggest that ^{239,240}Pu is moving more rapidly than a tracer (due to the smaller ratio between the measured concentrations in the TSA and LAVA). To obtain a match between model results and field data with a single scale factor for ^{239,240}Pu, a colloid velocity enhancement factor or smaller effective porosity is required. Shown in the figure is a fit to the ^{239,240}Pu breakthrough curve with an enhancement factor of $f_c = 1.06$ (compared to a maximum theoretical value of $f_c = 1.5$). This is equivalent to using an effective LAVA porosity of $0.012/1.06 = 0.0113$, and chimney porosity of $0.185/1.06 = 0.175$. Although these changes in porosity seem relatively minor, they can have a large effect on the travel time because of the small value of the porosity. Thus for a change $\Delta\phi$ in porosity, the travel time $\tau(\phi)$ becomes $\tau(\phi) \simeq \tau(\phi_0)(1 + \Delta\phi/\phi_0) = \tau(\phi_0)/f_c$.

It is reasonable to assume that the effective porosity used in the simulations, calibrated to the ^{3}H simulation, does not apply to colloids to explain the ^{239,240}Pu anomaly. In fact, the effective porosity experienced by a colloid should be less, because colloids generally do not participate in matrix diffusion as can conservative solutes. However, this does not explain why ^{137}Cs, also apparently transported irreversibly on colloids, does not exibit the more rapid migration experienced by ^{239,240}Pu. One possible explanation is that they are associated with different colloids with very different transport and radionuclide sorption properties. The species ^{137}Cs is a strongly sorbing cation which undergoes ion exchange and does not readily form complexes, whereas ^{239,240}Pu sorbs through surface complexation reactions and forms strong complexes with carbonate. Column transport experiments have been performed showing that ^{239,240}Pu sorbs nearly irreversibly zeolite colloids while sorbing reversibly on silica and clay colloids (Kersting and Reimus, 2003). Similar experiments for ^{137}Cs might elucidate the reversibility of that species on colloids and provide insight regarding the slightly different apparent velocities for ^{239,240}Pu and ^{137}Cs.

The scale factors in units of pCi/L obtained for each species by matching the breakthrough curve in the TSA to wells ER-20-5 #1 and #3 are listed in Table 2. The scale factors provide an estimate of the initial radionuclide inventory in Curies by multiplying by the melt-glass volume of 3.22×10^5 m^3 and porosity of 1%, to give the values listed in Table 2. It should be noted, however, that for the case of strongly sorbed species such as is expected for ^{137}Cs and ^{239,240}Pu, only the mobile species fraction that is irreversibly sorbed on colloids is accounted for in this estimate.

The simulation predicts a pulse release of radionuclides to the TSA and LAVA aquifers. Peak releases occur at roughly 15 y for the TSA unit and 55 y

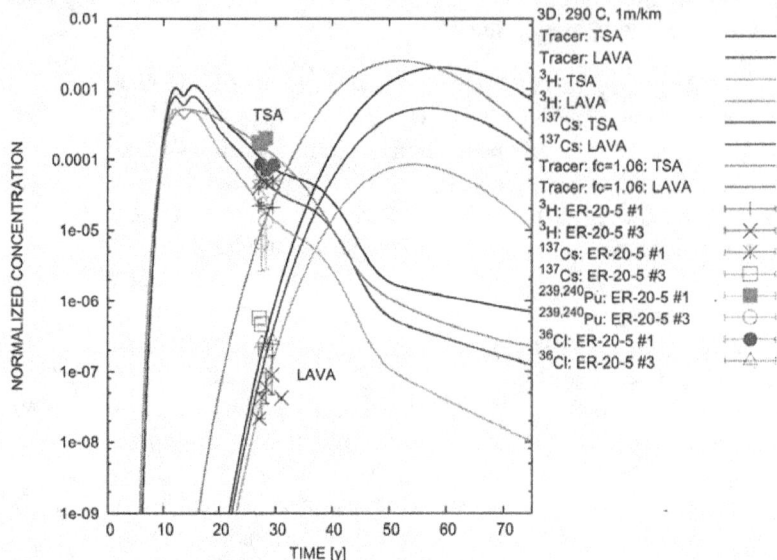

Figure 3: Breakthrough curves in the TSA and LAVA stratigraphic units for a 3-D simulation of a non-reactive tracer released from the melt glass. The breakthrough curves are corrected for radioactive decay of tritium and cesium (12.32 y and 30.2 y half lives). The measured ^{239,240}Pu breakthrough concentrations were fit with $f_c = 1.06$. Symbols represent field measurements from wells ER-20-5 #1 and #3 (Wolfsberg et al., 2002; Finnegan and Thompson, 2003), for the indicated radionuclide scaled by the factors listed in Table 2.

for the LAVA unit. This behavior is caused by the localized source terms for solute and heat in the initial melt glass configuration. Release of heat from the melt glass and formation of convection cells which drive fluid up the chimney, and its subsequent cooling and collapse of convective transport is, however, primarily responsible for the pulse behavior. Multicomponent, two-dimensional simulations of this system which include a kinetic model for dissolution of the melt glass and release of radionuclides also exhibit a pulse release through the temperature dependence of the dissolution rate which dramatically slows as the melt glass cools (Wolfsberg et al., 2002). The 3-D simulations suggest that tritium in well ER-20-5 #3 is just beginning to increase. However, the last ^{3}H observation point suggests that the peak in tritium has already occurred, marked by the observed drop in the tritium concentration in well ER-20-5 #3. This last data point was reported by Finnegan and Thompson (2003) for well ER-20-5 #3 (a pump failure occurred in well #1) taken on November, 2001. They measured ^{237}Np, ^{85}Kr and ^{3}H, and found no detectable ^{137}Cs. It should be noted, however, that the measurements from ER-20-5 #3 are close to the detection limit. Further data will be needed to confirm whether the model is correct and tritium and the other radionuclide concentrations are still increasing with time in well ER-20-5 #3.

There are a number of simplifying assumptions and uncertainties that may affect the results in this study. Fluxes through the aquifers are sensitive to the specified effective permeabilities. Single well tests indicate variations in permeability of several orders of magnitude within individual hydrostratigraphic units (Rehfeldt et al., 2004). Increasing the flux by increasing the head gradient to 2 m/km led to reversed order in concentrations with the larger value in the LAVA compared to the TSA and it was not possible to obtain a match to the field data. Additionally, fracture-matrix interactions (matrix diffusion in this case) are not accounted for explicitly. Extension to a dual continuum model formulation would enable incorporation of matrix diffusion, but this feature has not yet been incorporated into PFLOTRAN. The calculations do suggest, however, that the equivalent continuum representation of effective porosity is not valid. In this limiting case, the effective porosity is given by $\phi = \epsilon_f \phi_f + (1 - \epsilon_f)\phi_m \simeq \phi_m$, for $\epsilon_f \ll 1$. Considering that fracture porosities are very small and that the TSA and LAVA matrix porosities have the values 0.16 and 0.08 (Wolfsberg et al., 2002), respectively, this would give much larger values for the effective porosities than used in the simulation. Breakthrough times at ER-20-5 #1 in the TSA aquifer are rather insensitive to the TSA effective porosity provided it is sufficiently small because the travel time is dominated by flow through the chimney, but breakthrough at ER-20-5 #3 in the LAVA is very sensitive to the effective LAVA porosity, which is used as a fit parameter. However, these results suggest that in the TSA aquifer, because of its small effective porosity and fast travel time, matrix diffusion affects on non-reactive solutes and radionuclides irreversibly sorbed to colloids should be small. Finally, it should be noted that the initial distribution of radionuclides depends on the individual species (Pawloski et al., 2001). Radionuclides may be contained primarily in the melt glass (e.g. ^{239,240}Pu), or distributed throughout a much larger volume including the cavity and chimney but excluding the melt glass (e.g. ^{3}H), or be divided equally between the melt glass and cavity-chimney (e.g. ^{36}Cl). The particular distribution of each species would clearly affect the travel time to the TSA through the chimney.

Table 2: Scale factors used to fit measured radionuclide concentrations in well ER-20-5 #1 and #3 for a unit source concentration and the estimated mobile fraction of the initial inventory.

Radionuclide	Scale Factor [pCi/L]	Estimated Inventory [Ci]
^{3}H	3×10^{12}	9.66×10^{6}
^{36}Cl	4×10^{4}	0.129
^{137}Cs*	1×10^{6}	3.22
^{239,240}Pu*	3×10^{3}	0.01

* Mobile fraction irreversibly sorbed on colloids.

3 CONCLUSION

Three-dimensional simulations at the field scale demonstrate the feasibility of matching field data documenting radionuclide migration from the BENHAM underground nuclear test and thereby narrowing the range of uncertainty in model simulations. The model simulations show a pulse release of radionuclides to the TSA aquifer over a narrow window in time through the creation of convection cells caused by heat released from the melt glass. Field data from wells ER-20-5 #1 and #3 for radionuclides ^{3}H, ^{36}Cl, and ^{137}Cs could be explained assuming a non-reactive tracer with concentrations adjusted for radioactive decay, although additional data is needed to confirm the trend in time. Through the use of effective properties for porosity and permeability, allowance could be made for matrix diffusion processes, the extent of which may be limited by high flow velocities and possibly fracture coatings that inhibit matrix diffusion. The radionuclide ^{239,240}Pu presented an anomaly in the fit to the field data, which indicated it was moving faster than the other three radionucldes considered. While this is consistent with the hypothesis that ^{239,240}Pu transport is colloid-facilitated, because the fit to the ^{137}Cs field data also indicated colloid-facillitated transport but at somewhat slower migration rates than ^{239,240}Pu, presumably different types of colloids were involved for ^{239,240}Pu and ^{137}Cs. The results presented here appear to be the first indication that field scale colloid transport can involve flow velocities faster than a non-reactive tracer species.

Because simulations of radionuclide migration from an underground nuclear test, such as the BENHAM test, are intrinsically 3-D as a result of the unique cavity-chimney-melt-glass geometry, massively parallel high performance computing is required to tackle these problems. Using the computer code PFLO-TRAN it is now possible to carry out such calculations on available computing facilities such as LANL's QSC nonclassified alpha cluster, as well as other machines including EMSL's MPP2 Intel Itanium-2 cluster at PNNL. However, it should perhaps be emphasized that high performance computing can help alleviate, but not eliminate, upscaling and multiscale issues. However, high performance computing will make more feasible multiple continuum models which can incorporate length scales ranging from perhaps millimeters to meters for describing multicomponent-multiphase systems in fractured porous media.

Future work will consider the relatively immobile radionuclide fraction through the addition of multicomponent chemical reactions including reversible and irreversible sorption, complexation, and precipitation and dissolution reactions to describe alteration of the melt glass, and implementation of a dual continuum description of fracture-matrix interaction. The effect of grid refinement on convection cells formed during the cooling period will be investigated. Finally, the rewetting phase of the near-field region and its effect on radionuclide release times will be considered.

4   ACKNOWLEDGEMENTS

We would like to thank Robert M. Bangerter, NNSA Underground Test Area (UGTA) manager, for supporting our initial studies (Wolfsberg et al., 2002), and for review and advice regarding this manuscript. Hari Viswanathan provided a helpful review of an earlier version of the manuscript. We also thank Richard Mills and Glenn Hammond for help with implementing parallel computing using PETSc. This research was performed in part using the MSCF in EMSL, a national scientific user facility sponsored by the U.S. DOE, OBER and located at PNNL, and the QSC machine at LANL. This work was funded in part by the Los Alamos National Laboratory LDRD-DR project 20030091DR.

5   REFERENCES

Balay, S., W.D. Gropp, L.C. McInnes and B.F. Smith. 1997. Efficient Management of Parallelism in Object Oriented Numerical Software Libraries, Modern Software Tools in Scientific Computing, E.Arge, A.M. Bruaset and H.P. Langtangen, Ed., Birkhauser Press, Boston, MA, 163-202.

Department of Energy. 2000. United States Nuclear Tests-July 1945 through September 1992. US Department of Energy, Nevada Operations Office, Las Vegas, NV, DOE/NV-209 (Rev. 15).

Honeyman, B. D., and J. F. Ranville. 2002. Colloid properties and their effects on radionuclide transport through soils and groundwater, in Geochemistry of Soil Radionuclides, edited by P.-C. Zhang, and P.V. Brady, pp. 131–163, Soil Science Society of America, Special Publication Nr. 59, Madison, WI.

Finnegan, D.L. and J.L. Thompson. 2003. Laboratory and Field Studies Related to Radionuclide Migration at the Nevada Test Site in Support of the Underground Test Area Project and Hydrologic Resources Management Project, October 1, 2001 - September 30, 2002, Los Alamos National Laboratory LA-14042-PR, Los Alamos, NM.

Kersting, A.B., D.W. Efurd, D.L. Finnegan, D.J. Rokop, D.K. Smith, and J.L. Thompson. 1999. Migration of plutonium in groundwater at the Nevada Test Site, Nature 397: 56–59.

Kersting, A.B., and P.W. Reimus. 2003. Colloid-Facilitated Transport of Low-Solubility Radionuclides: A Field, Experimental, and Modeling Investigation, UCRL-ID-149688. Lawrence Livermore National Laboratory, Livermore, CA; and Los Alamos National Laboratory, Los Alamos, NM.

Lichtner, P.C. 2000. Critique of Dual Continuum Formulations of Multicomponent Reactive Transport in Fractured Porous Media, *Dynamics of Fluids in Fractured Rock*, Geophysical Monograph **122**, 281–298.

Pawloski, G.A. 1999. Development of phenomenological models of underground nuclear tests on Pahute Mesa, Nevada Test SiteBENHAM and TYBO, Lawrence Livermore National Laboratory, UCRL-ID-136003.

Pawloski, G.A., A.F.B. Tompson, and S.F. Carle (eds.). 2001. Evaluation of the hydrologic source term from underground nuclear tests on Pahute Mesa at the Nevada Test Site: The Cheshire test, Lawrence Livermore National Laboratory, UCRL-ID-147023.

Rehfeldt, K., W. Drici, B. Lester, D. Sloop, J. Watrus, T.Beard, M. Sully, W. Fryer, and C. Benedict. 2004. Hydrologic Data for the Groundwater Flow and Contaminant Transport Model of Corrective Action Units 101 and 102: Central and Western Pahute Mesa, Nye County, Nevada. Technical report S-N/99205–002, Stoller-Navaro, Las Vegas Nevada.

Wolfsberg, A., L. Glascoe, G. Lu, A. Olson, P. Lichtner, M. McGraw, T. Cherry, and G. Roemer. 2002. TYBO/BENHAM: Model Analysis of Groundwater Flow and Radionuclide Migration from Underground Nuclear Tests in Southwestern Pahute Mesa, Nevada, Los Alamos National Laboratory LA-13977, Los Alamos, NM.

4.23 Challenges in modeling evolving reactivity: Examples from groundwater contamination and remediation studies

K.U. Mayer[1], T. H. Henderson[1], S.-W. Jeen[2], D. W. Blowes[2], B. L. Parker[2], T. A. Al[3]., and R. W. Gillham[2]

[1] *University of British Columbia, Vancouver, British Columbia, Canada, e-mail: umayer@eos.ubc.ca.*

[2] *University of Waterloo, Waterloo, Ontario, Canada.*

[2] *University of New Brunswick, Fredericton, New Brunswick, Canada.*

1 INTRODUCTION

Significant advances have been made in recent years with respect to the modeling of complex reactive transport systems. Increased computer resources now allow the simulation of two- and three-dimensional problems involving a large number of unknowns and various geochemical processes. The parameter demand for these simulations is significant, but numerical models have nevertheless proven useful to test and investigate conceptual models in a quantitative manner, provided adequate aqueous, gaseous and mineralogical data is available. One of the remaining challenges focuses on the issue of evolving reactivity, which can manifest itself in changing (bio)-geochemical reaction rates or changes in the mixing characteristics of pore water and/or NAPL's within an aquifer. If part of the objective of a modeling study is to investigate the evolution of a system, evolving reactivity often needs to be taken into account. Here, we utilize two examples involving the remediation of contaminated groundwater to highlight the importance of the consideration of evolving reactivity, when modeling long-term system evolution. All simulations presented were conducted using the multicomponent reactive transport model MIN3P (Mayer et al., 2002).

2 PERMEABLE REACTIVE BARRIERS

2.1 *Elizabeth City Site – North Carolina*

Multicomponent reactive transport modeling was conducted for the permeable reactive barrier at the Coast Guard Support Center near Elizabeth City, N.C (Mayer et al., 2001). The zero-valent iron barrier was installed to treat groundwater contaminated by hexavalent chromium and chlorinated solvents (Blowes at al., 1999). Among the simulated processes were contaminant destruction and the reduction of other electron acceptors causing the consumption of the treatment material, and the precipitation of secondary minerals.

2.2 *Simulation Results*

The model results were able to illustrate the removal of Cr(VI) and the chlorinated solvents by the reactive barrier and demonstrate that reactions other than the remediation reactions most significantly affect the water chemistry in the barrier. In particular sulfate reduction and iron corrosion appear to control the evolution of the pore water while passing through the treatment system. The associated pH increase causes the precipitation of significant amounts of carbonate, hydroxide and sulfide mineral phases (Figure 1). The original simulation results indicate that secondary mineral formation has the potential to decrease the porosity in the barrier over the long term and suggest that the precipitation of minerals is concentrated in the up-gradient portion of the barrier (Figure 1). Decreasing iron reactivity due to passivation was not accounted for in this study. However, it was hypothesized that passivation causes less depletion of the treatment material and precipitation of secondary minerals in the inflow area than simulated (Mayer et al., 2001).

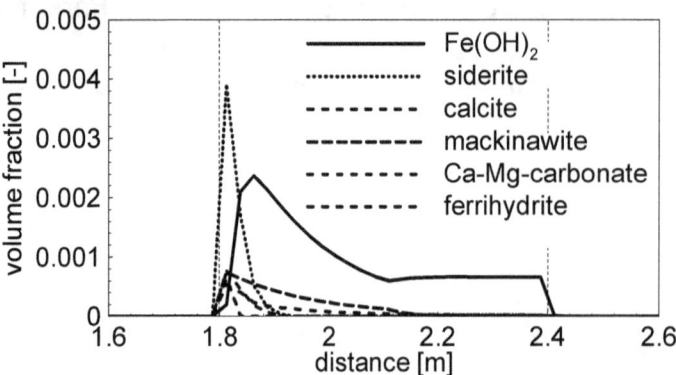

Figure 1: Simulated volume fractions of secondary minerals in reactive barrier after 240 days

2.3 *Comparison to Field Observations*

The simulations presented above have been continued to 4 years of barrier operation and the accumulated inorganic carbon (Figure 2) and sulfur (not shown) has been compared to measured accumulations extracted from cores (Wilkin et al., 2002). Although observed and simulated results compare well in general, the observed data shows a near continuous mass distribution of precipitate within the first 5-6 cm of the barrier, while the simulated results show a continuous decrease in precipitate accumulation along the flowpath. The observed data indicate that some passivation may have taken place, which was not accounted for by the model.

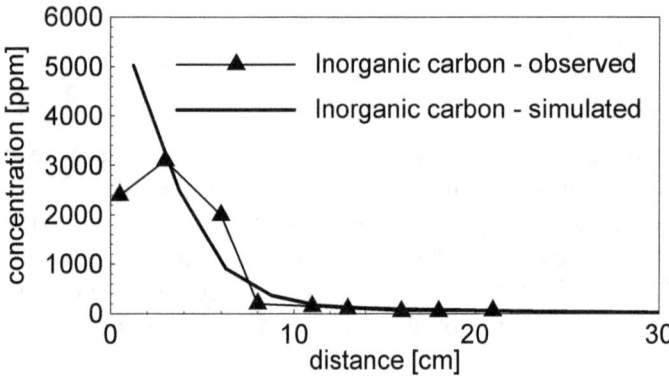

Figure 2: Simulated and observed inorganic carbon contents after 4 years (Data from Wilkin et al., 2002)

2.4 *Long term column studies*

Currently, column studies are conducted at the University of Waterloo (Jeen, Blowes and Gillham, work in progress), which confirm the hypothesis that passivation due to carbonate mineral precipitation is occurring. These observations highlight the need to account for evolving reactivity in reactive transport models. Other processes that may affect system reactivity in permeable reactive barriers include the formation of hydrogen gas bubbles, or the depletion of the more reactive portion of treatment materials (e.g.: Herbert et al., 2000).

3 IN SITU CHEMICAL OXIDATION USING PERMANGANATE

3.1 *Connecticut Site*

Three-dimensional simulations of permanganate oxidation of chlorinated solvents are being conducted for a site located near Hartford, Connecticut. At this site, in-situ chemical oxidation of the DNAPL using potassium permanganate was tested over a seven month period in 2000 and 2001 (Stewart, 2002). The oxidant delivery methodology was based on the inject-and-leave method (Nelson et al., 2000) and relied on two sequential injections of potassium permanganate solutions with concentrations approaching aqueous solubility. The simulations account for the movement of variable density fluids, water table mounding, and evolving aqueous permeability due to the dissolution of non-wetting fluids and the dissolution/precipitation of mineral phases. The geochemical reaction network includes kinetic expressions describing dissolution and precipitation of mineral phases, DNAPL dissolution, and the intra-aqueous oxidation reaction between permanganate and dissolved TCE. Mineral phases in the simulations include calcite and manganese oxides, and organic carbon is used to simulate the oxidant demand of the aquifer. The spread of the oxidant, and the reaction between the oxidant and TCE was monitored over time using an array of multilevel monitoring points (Stewart 2002), which is available for comparison with the simulated results.

3.2 *Simulation Results*

To illustrate the problem, simulated data for aqueous permanganate, and TCE are shown in Figure 3. An initial evaluation of model performance is being performed by comparing observed and simulated data for aqueous permanganate, chloride, and dissolved TCE. The data indicate the model accurately depicts the lateral extent of the spread of permanganate, and generally predicted aqueous chloride concentrations within a factor of 2 of the field data. However, the permanganate data indicated that the model generally under-predicting the persistence of the permanganate in the immediate vicinity of the aquitard contact, where NAPL is present.

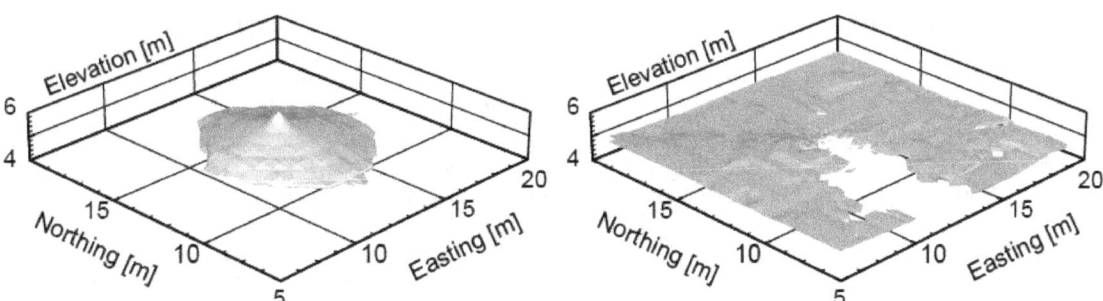

Figure 3: 1 g/L permangante iso-contour surface (left) and 1 percent DNAPL volume fraction iso-contour surface (right) at 14 days following initial permanganate injection. Injection point location was 12.5 m northing, 12.5 m easting, and elevation 5.6 meters

There are several factors that may affect persistence of permanganate, or in other words system reactivity, most of which are related to the precipitation of manganese oxides. MnO_2-precipitation may result in a permeability decrease, which may be counteracted by carbonate mineral dissolution. The precipitation of manganese oxides may also affect system reactivity (limiting the dissolution of both NAPL and carbonate minerals). These results suggest that at least some of these processes need to be included in the model formulation to allow a more realistic simulation of in-situ chemical oxidation.

4 CONCLUSIONS

Examples from groundwater remediation are particularly useful to illustrate the effects of evolving reactivity due to rapid reaction rates, a high degree of disequilibrium, steep geochemical gradients, and pronounced geochemical changes. However, many of these conclusions are also valid for large scale reactive transport studies, even in pristine aquifers. Processes of importance include decreases in reactivity due to the formation of surface coatings and passivation (Wunderly et al., 1996, Mayer et al., 2002), increase in reactivity due to generation of surface sites (e.g.: Postma and Appelo, 1998), but also permeability changes due to changes in aqueous phase saturation as a result of NAPL dissolution or gas formation, which may result in hydraulic exclusion. Some of these processes have been considered previously in reactive transport models, but to date have not been included in a comprehensive manner. This limits the applicability of reactive transport models to study system evolution and to use these models in a predictive sense. Reasons for the lack of implementing evolving reactivity include the inherent complexity of the interacting processes, the difficulty of quantifying the processes that control reactivity, and their often site-specific nature.

REFERENCES

Blowes, D. W., Gillham, R. W., Ptacek, C. J., Puls, R. W., Bennett, T. A., Bain, J. G., Hanton-Fong, C. J., andPaul, C. J., 1999. An in-situ permeable reactive barrier for the treatment of hexavalent chromium and trichloroethylene in groundwater: Vol. 2: Performance monitoring, U. S. Environmental Protection Agency, EPA/600/R-99/095b.

Herbert, R.B., Benner, S.G. and Blowes, D.W. 2000, Solid phase iron - sulfur geochemistry of a reactive barrier for treatment of mine drainage, *Appl. Geochem.*. 15, 1331 - 1343.

Mayer, K. U., D. W. Blowes, and E. O. Frind, 2001. Reactive transport modelling of groundwater remediation by an in-situ reactive barrier for the treatment of hexavalent chromium and trichloroethylene, *Water Resour. Res.*, 37:3091-3103.

Mayer, K. U., E. O. Frind, and D. W. Blowes, 2002. Multicomponent Reactive Transport Modeling in Variably Saturated Porous Media Using a Generalized Formulation for Kinetically Controlled Reactions, *Water Resour. Res.*, 38, doi: 10:1029/2001WR000862.

Nelson, M.D., Parker, B.L., Al, T.A., Cherry, J.A. and Loomer, D. 2001. Geochemical reactions resulting from *in situ* oxidation of PCE DNAPL by $KMnO_4$ in a sandy aquifer. *Env. Sci. Technol.*, 35:1266-1275.

Postma, D, and Appelo, C. A. J., 2000. Reduction of Mn-oxides by ferrous iron in a flow system: column experiment and reactive transport modeling, *Geochim. Cosmochim, Acta*, 64:1237-1247.

Stewart, C., 2002. Density-driven permanganate solution delivery and chemical oxidation of a thin trichloroethene DNAPL pool in a sandy aquifer, M.Sc Thesis, University of Waterloo, Ontario, Canada.

Wilkin, R. T., Puls, R. W., and Sewell, G. W., 2002. Long term performance of permeable reactiove barriers using zero-valent iron: An evaluation at two sites. U.S. Environmental Protection Agency Research Brief. EPA/600/S-02/001.

Wunderly, M. D., Blowes, D. W. and Frind, E. O. and Ptacek, C. J., 1996. Sulfide mineral oxidation and subsequent reactive transport of oxidation products in mine tailings impoundments: A numerical model, *Water Resour. Res.*, 32:3173-3187

5. WORKSHOP PARTICIPANTS

Conceptual Model Development for Subsurface Reactive Transport Modeling of Inorganic Contaminants, Radionuclides, and Nutrients

La Posada de Albuquerque
Albuquerque, New Mexico
April 19-22, 2004

Aelion, C. Marjorie	University of South Carolina	aelionm@sc.edu
Altman, Susan	Sandia National Laboratories	sjaltma@sandia.gov
Appelo, Cornelis A. J.	Hydrochemical, The Netherlands	appt@xs4all.nl
Apps, John A.	Lawrence Berkeley National Laboratory	apps@lbl.gov
Balistrieri, Laurie S.	U.S. Geological Survey	balistri@usgs.gov
Ball, William P.	Johns Hopkins University	bball@jhu.edu
Beckie, Roger	University of British Columbia, Canada	rbeckie@eos.ubc.ca
Bertetti, Paul	Southwest Research Institute	pbertetti@swri.edu
Blowes, David	University of Waterloo, Canada	blowes@uwaterloo.ca
Blum, Alex	U.S. Geological Survey	aeblum@usgs.gov
Bodvarsson, Gudmundr (Bo)	Lawrence Berkeley National Laboratory	gsbodvarsson@lbl.gov
Brantley, Susan L.	Pennsylvania State University	brantley@essc.psu.edu
Burgos, Bill	Pennsylvania State University	wdb3@psu.edu
Cantrell, Kirk	Pacific Northwest National Laboratory	kirk.cantrell@pnl.gov
Criscenti, Louise	Sandia National Laboratories	jcrisc@sandia.gov
Curtis, Gary	U.S. Geological Survey	gpcurtis@usgs.gov
Cygan, Randall T.	Sandia National Laboratories	rtcygan@sandia.gov
Davis, James A.	U.S. Geological Survey	jadavis@usgs.gov
Destouni, Georgia	Stockholm University, Sweden	georgia.destouni@natgeo.su.se
Dyer, James A.	DuPont Engineering Technology	james.a.dyer@usa.dupont.com
Ford, Robert G.	U.S. Environmnetal Protection Agency	ford.robert@epa.gov
Fryberger, Teresa	U.S. Department of Energy	teresa.fryberger@science.doe.gov
Gao, Suduan	USDA-ARS Water Management Research Lab	sugao@ucdavis.edu
Ginn, Timothy R.	University of California at Davis	trginn@ucdavis.edu
Goldberg, Sabine R.	USDA-ARS Salinity Laboratory	sgoldberg@ussl.ars.usda.gov
Gwo, Jin-Ping	University of Maryland	jgwo@umbc.edu
Hammond, Glenn	Sandia National Laboratories	gehammo@sandia.gov
Harvey, Charles F.	Massachusetts Institute of Technology	charvey@mit.edu
Hering, Janet	California Institute of Technology	hering@caltech.edu
Herman, Janet S.	University of Virginia	jherman@virginia.edu
Hill, Mary	U.S. Geological Survey	mchill@usgs.gov
Hunt, James R.	University of California at Berkeley	hunt@ce.berkeley.edu
Honeyman, Bruce D.	Colorado School of Mines	honeyman@mines.edu
Jove-Colon, Carlos F.	Sandia National Laboratories	cfjovec@sandia.gov
Kent, Douglas	U.S. Geological Survey	dbkent@usgs.gov
Koretsky, Carla	Western Michigan University	carla.koretsky@wmich.edu

Krupka, Kenneth M.	Pacific Northwest National Laboratory	ken.krupka@pnl.gov
Lichtner, Peter C.	Los Alamos National Laboratory	lichtner@lanl.gov
McGrath, Chris	US Army Engineer R&D Center	chris.mcgrath@erdc.usace.army.mil
Mayer, Ulrich	University of British Columbia, Canada	umayer@eos.ubc.ca
Meigs, Lucy C.	Sandia National Laboratories	meigs@swcp.com
O'Day, Peggy	University of California at Merced	poday@ucmerced.edu
Oates, Peter	Massachusetts Institute of Technology	pproof@aol.com
Ott, William	U.S. Nuclear Regulatory Commission	wro1@nrc.gov
Pabalan, Roberto	Southwest Research Institute	rpabalan@cnwra.swri.edu
Painter, Scott	Southwest Research Institute	spainter@swri.org
Palmer, Carl D.	Idaho National Laboratory	palmcd@inel.gov
Parkhurst, David	U.S. Geological Survey	dlpark@usgs.gov
Pichler, Thomas	University of South Florida	pichler@chuma.cas.usf.edu
Puls, Robert	U.S. Environmnetal Protection Agency	puls.robert@epa.gov
Redden, George D.	Idaho National Laboratory	reddgd@inel.gov
Reddy, Michael M.	U.S. Geological Survey	mmreddy@usgs.gov
Ridge, A. Christianne	U.S. Nuclear Regulatory Commission	acr1@nrc.gov
Roden, Eric E.	University of Alabama	eroden@bsc.as.ua.edu
Sahai, Nita	University of Wisconsin at Madison	sahai@geology.wisc.edu
Samper, F. Javier	University of La Coruna, Spain	jsc@iccp.udc.es
Schreiber, Madeline	Virginia Tech	mschre b@vt.edu
Siegel, Malcolm	Sandia National Laboratories	msiegel@sandia.gov
Sonnenthal, Eric	Lawrence Berkeley National Laboratory	ersonnenthal@lbl.gov
Steefel, Carl I.	Lawrence Berkeley National Laboratory	CIsteefel@lbl.gov
Stevens, Caroline	U.S. Environmnetal Protection Agency	stevens.caroline@epa.gov
Stillings, Lisa	U.S. Geological Survey	stilling@usgs.gov
Su, Chunming	U.S. Environmnetal Protection Agency	su.chunming@epa.gov
Suarez, Donald L.	USDA-ARS Salinity Laboratory	dsuarez@ussl.ars.usda.gov
Tompson, Andrew F. B.	Lawrence Livermore National Laboratory	afbt@llnl.gov
Turner, David	Southwest Research Institute	dturner@swri.org
Valocchi, Albert J.	University of Illinois	valocchi@uiuc.edu
Van Cappellen, Philippe, S.	Utrecht University, The Netherlands	pvc@geo.uu.nl
Warren, Lesley A.	McMaster University, Canada	warrenl@mcmaster.ca
Westall, John C.	Oregon State University	john.westall@oregonstate.edu
Yabusaki, Steve	Pacific Northwest National Laboratory	yabusaki@pnl.gov
Yeh, Gour-Tsyh	University of Central Florida	gyeh@mail.ucf.edu
Wang, Yifeng	Sandia National Laboratories	ywang@sandia.gov
Zachara, John	Pacific Northwest National Laboratory	john.zachara@pnl.gov